TECHNICAL REPORT

Explaining the Increase in Unemployment Compensation for Ex-Servicemembers During the Global War on Terror

David S. Loughran, Jacob Alex Klerman

Prepared for the Office of the Secretary of Defense

NATIONAL DEFENSE RESEARCH INSTITUTE

The research described in this report was prepared for the Office of the Secretary of Defense-Reserve Affairs and conducted within the Forces and Resources Policy Center of the RAND National Defense Research Institute (NDRI), a federally funded research and development center sponsored by the Office of the Secretary of Defense, the Joint Staff, the Unified Combatant Commands, the Department of the Navy, the Marine Corps, the defense agencies, and the defense Intelligence Community under Contract W74V8H-06-C-0002.

Library of Congress Cataloging-in-Publication Data is available for this publication.

ISBN 978-0-8330-4514-0

Published 2008 by the RAND Corporation
1776 Main Street, P.O. Box 2138, Santa Monica, CA 90407-2138
1200 South Hayes Street, Arlington, VA 22202-5050
4570 Fifth Avenue, Suite 600, Pittsburgh, PA 15213-2665
RAND URL: http://www.rand.org/
To order RAND documents or to obtain additional information, contact
Distribution Services: Telephone: (310) 451-7002;
Fax: (310) 451-6915; Email: order@rand.org

Preface

This document was produced as part of the RAND project "Understanding Recent Trends in Veteran Unemployment." That project was motivated by increases in the official unemployment rate for young veterans and in the number of veterans claiming Unemployment Compensation for Ex-Servicemembers (UCX), both of which raised concern that recent veterans were experiencing trouble transitioning from military service to the civilian labor market. Earlier research under this project concluded that the increase in the official unemployment rate of young veterans is not cause for concern by itself (Savych, Klerman, and Loughran, 2008). But the sharp increase in the UCX caseload is nonetheless worrisome. In the research reported here, we examine the reasons why the UCX caseload has increased and discuss the implications of those findings for the UCX program. This report will be of interest to policymakers and manpower analysts concerned about the transition of active- and reserve-component members from active-duty service to civilian jobs and the development of programs designed to facilitate that transition.

This research was sponsored by the Office of the Secretary of Defense for Reserve Affairs and conducted within the Forces and Resources Policy Center of the RAND National Defense Research Institute (NDRI), a federally funded research and development center sponsored by the Office of the Secretary of Defense, the Joint Staff, the Unified Combatant Commands, the Department of the Navy, the Marine Corps, the defense agencies, and the defense intelligence community.

Comments regarding this document are welcome and may be addressed to David Loughran by email at david_loughran@rand.org. For more information about RAND's Forces and Resources Policy Center, contact the Director, James Hosek, by email at james_hosek@rand.org. Loughran and Hosek can be reached by phone at 310-393-0411 or by mail at the RAND Corporation, 1776 Main Street, Santa Monica, California 90407-2138. More information about RAND is available at http://www.rand.org.

Contents

Figures

Tables

Summary

Between 2002 and 2004, the number of veterans receiving Unemployment Compensation for Ex-Servicemembers and the cost of this program to the U.S. Department of Defense (DoD) increased by about 75 percent (see Figure S.1). The UCX program is the military counterpart to the civilian Unemployment Insurance (UI) program, which provides income assistance to the unemployed as they search for work. Honorably discharged active-component personnel and reserve-component personnel completing a period of active-duty service of 90 or more days are eligible to receive UCX benefits provided that they meet other federal and state-specific requirements of the UI system.

The sharp and sustained increase in the UCX caseload since 2002 has contributed to concerns that veterans of the wars in Iraq and Afghanistan are having difficulty transition-

Figure S.1
Average Weekly UCX Caseload, by Quarter

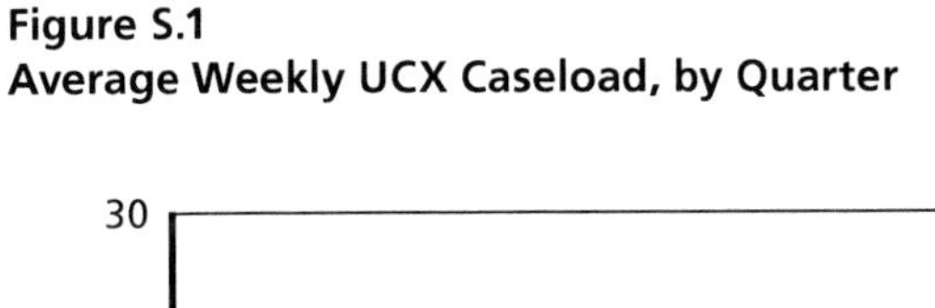
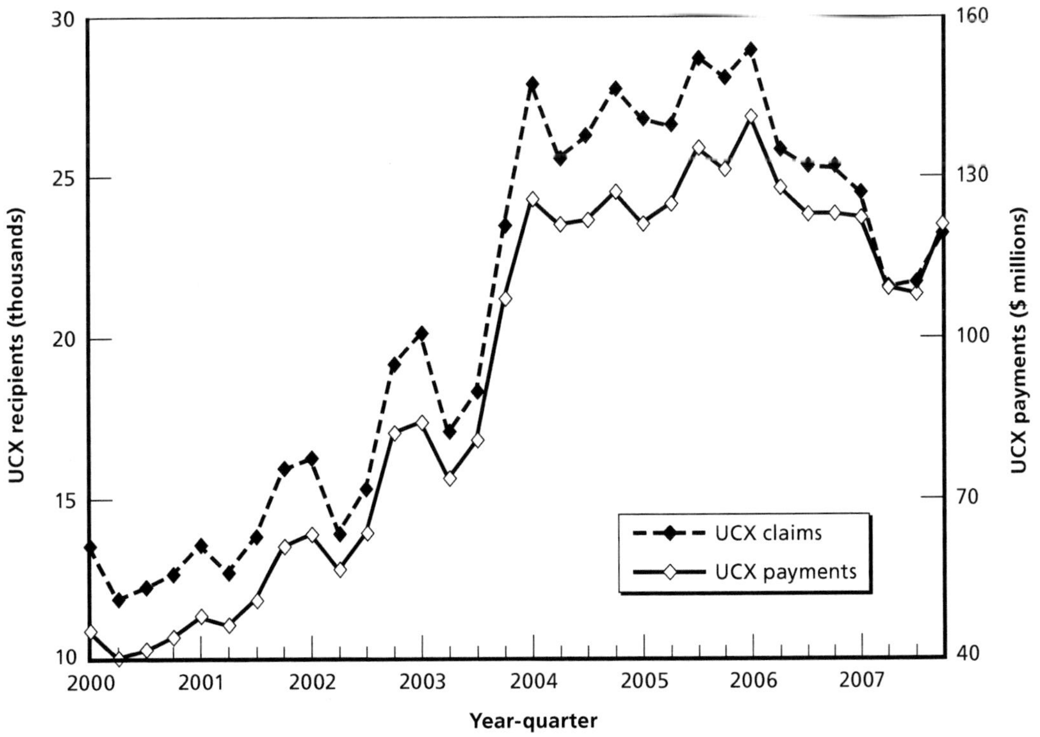

SOURCE: U.S. Department of Labor, Employment and Training Administration.
RAND *TR588-S.1*

ing to the civilian labor market. The research reported in this document examines the reasons why the UCX caseload has risen and considers the implications of those findings for the UCX program.

Data

To understand this increase in the UCX caseload, we drew on several data sources. We began by analyzing aggregate data on the UCX program maintained by the U.S. Department of Labor (DoL), Employment and Training Administration (ETA). We then created an individual-level database of UCX recipients. That database began with a sample of UCX recipients obtained from hard-copy reports maintained by the individual military services. Data limitations and considerations of cost led us to select two quarters of these data to be key-punched into electronic form: the second quarter of 2005 for the Army, Air Force, Navy, and Marine Corps and the second quarter of 2002 for the Army and Air Force alone (data on the Navy and Marine Corps were not available for that quarter). The two quarters were selected so as to span the observed increase in the UCX caseload.

We augmented these individual-level data on UCX claims by incorporating information from service records on component, rank, dates of service and deployment, and self-reported health from the Post-Deployment Health Assessment (PDHA). We also analyzed survey data on post-activation employment of reservists from the Status of Forces Survey of Reserve Component Members (SOFS-R).

Rising UCX Eligibility and Claim Rates

We find that the increase in the UCX caseload is attributable both to large increases in the number of veterans potentially eligible to receive UCX and to large increases in the fraction of potentially eligible veterans who claim UCX (the claim rate). As can be seen in Figure S.2, the number of reservists completing an activation of 90 or more days, and therefore potentially eligible to receive UCX, increased sharply following September 11, 2001, as reservists were deployed to assist with homeland security and to support operations in Afghanistan. With the beginning of Operation Iraqi Freedom in early 2003, reserve deactivations increased to unprecedented levels. By contrast, active-component separations have been relatively stable since 2000.

These trends in the numbers of veterans eligible to receive UCX suggest that a significant proportion of the overall increase in the UCX caseload is attributable to the intensive use of the reserve components in the Global War on Terror. The individual-level data allow us to identify the service and component of UCX claimants. Computations based on these individual UCX claim data for the second quarter of 2002 and the second quarter of 2005 imply that approximately 58 percent of the increase in the UCX caseload between 2002 and 2005 is attributable to the Army reserve components. We estimate that the Air Force, Navy, and Marine Corps reserve components account for a very small fraction of the overall increase in the UCX caseload between those years, which leaves the balance of the increase (about 40 percent) to the active components.

Figure S.2
Active- and Reserve-Component Demobilizations, by Quarter

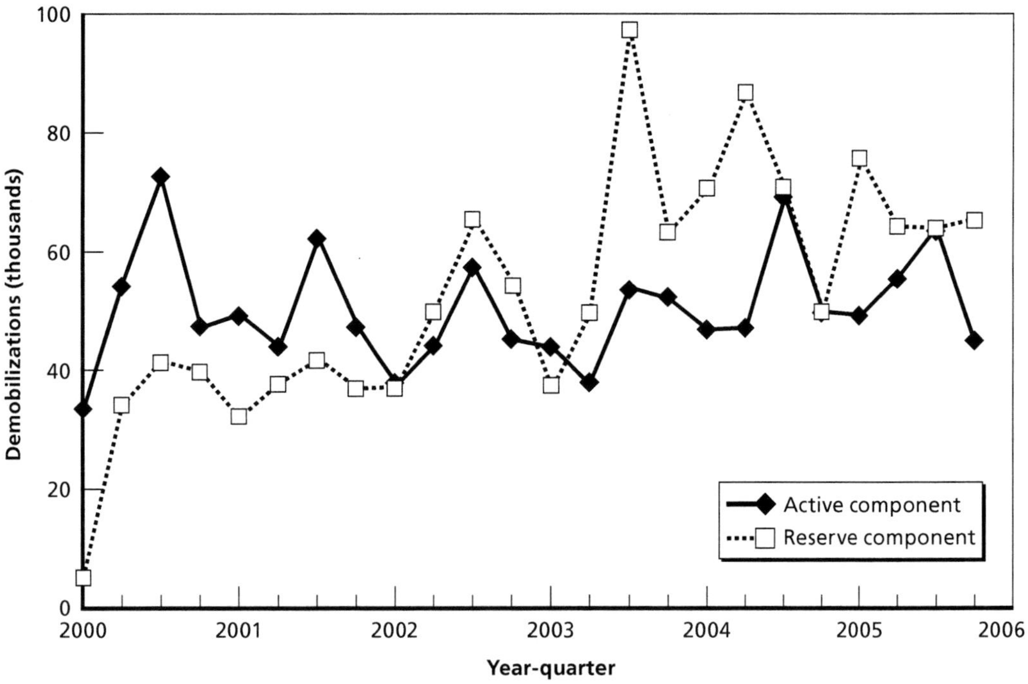

SOURCES: Work Experience file, Active Duty Pay file, and Reserve Pay file.
NOTE: Counts exclude the Coast Guard.
RAND *TR588-S.2*

We also use these individual-level data to decompose the increase in the Army and Air Force UCX caseload into two parts: (1) increases in eligibility and (2) increases in the fraction of eligible reservists who claim UCX (data limitations prevent us from performing this decomposition for the Navy and Marine Corps). Specifically, from the individual-level data, we compute the probability of claiming UCX in a quarter for each component. Then, we compute the effect of changes in eligibility by using observed eligibility in each period, but holding the claim rates constant at their mean value. Similarly, we compute the effect of changes in claiming by using observed claiming rates, but holding the number of eligible servicemembers constant at their mean value.

Table S.1 tabulates the result of that analysis. Eligibility more than doubled in the Air Force reserve components and nearly tripled in the Army components. Changes in eligibility explain more than half the increase in the UCX caseload in both the Army and Air Force reserve components. Changes in active-component eligibility are much smaller (14 percent in the Army and 12 percent in the Air Force). Nevertheless, changes in eligibility explain about

Table S.1
The Effect of Eligibility and Claiming on the Increase in the UCX Caseload Between the Second Quarters of 2002 and 2005

	Active Component		Reserve Component	
	Army	Air Force	Army	Air Force
Percentage increase in eligibility	14	12	175	106
Percentage-point increase in claim rate	0.6	2.5	4.9	0.9
Share of increase in UCX caseload attributable to				
Increase in eligibility	26	55	54	53
Increase in claim rates	74	45	46	47

SOURCE: RAND UCX database.

one-quarter of the increase in the UCX caseload in the Army active component and more than half of the increase in the Air Force active component.

Deployment Duration, Post-Deployment Health, and UCX Claim Rates

While increases in the number of veterans eligible to receive UCX have been important, rising claim rates have also played an important role. Nearly half of the increase in the Air Force active- and reserve-component caseload, three-quarters of the Army active-component caseload, and half of the Army reserve-component caseload are accounted for by rising claim rates.

It seems unlikely that the increase in claim rates is due to a deterioration in the civilian labor market. Between 2002 and 2005, the overall civilian labor market generally improved; for example, civilian UI claims decreased markedly between these two years.[1] Instead, it seems likely that the increase in claim rates is due to changes in the nature of military service. We show that long deployments have become much more common and that longer deployments are associated with much higher claim rates. Our decomposition analysis suggests that these longer deployments explain more than a third of the overall increase in the Army active and reserve UCX caseload between 2002 and 2005 (similar analyses could not be conducted for other services).

We also explored the relationship between post-deployment health status (as measured in the PDHA) and UCX utilization among Army active- and reserve-component members. We show that self-reported health worsens with length of deployment and that poor reported health is associated with higher UCX claim rates. However, while these correlations are statis-

[1] The official Current Population Survey (CPS-based unemployment rate of younger veterans (ages 20–24) increased between 2003 and 2005 (while the unemployment rate of nonveteran youth declined). However, the official unemployment rate returned to 2003 levels in 2006. The decline in the unemployment rate of younger veterans in 2006 and a more formal analysis in earlier research (Savych, Klerman, and Loughran, 2008) suggest that the observed increase in veteran youth unemployment between 2003 and 2005 most likely reflects sampling variation rather than the influence of real economic factors. Thus, there is no clear evidence of either an improvement or worsening of the employment prospects of young veterans.

tically significant, our analysis suggests that they are not large enough to explain much of the overall increase in the Army UCX caseload.

The Employment Experiences of Army Reserve-Component UCX Recipients

In 2005, more than 19 percent of eligible Army active-component and 15 percent of eligible Army reserve-component members claimed UCX benefits following demobilization. The high claim rates of Army active-component members are perhaps more easily understood than the high claim rates of Army reserve-component members. Most active-component veterans have no recent civilian labor market experience. It is, therefore, plausible that many will require some time to find civilian employment that takes advantage of the skills they developed in the military.

The situation with reservists is quite different. The Uniformed Services Employment and Reemployment Rights Act (USERRA) of 1994 (P. L. 103-353) guarantees reservists employed at the time of activation the right to return to that job following activation. Since, as with UI, UCX generally requires that recipients accept suitable employment, the availability of a USERRA-protected job should make it less likely that a reservist would be eligible to receive UCX.[2]

However, our tabulations from the SOFS-R indicate that 59 percent of reservists who received UCX in the three months following deactivation were employed in the month prior to being activated. The SOFS-R asked these reservists why they had not returned to their pre-activation job. We classified those reasons as either involuntary (e.g., employer went out of business, change in employer circumstances, failure to offer prompt employment) or voluntary (e.g., disliked previous job, decided to attend school, "needed a break") in nature. Overall, among Army reservists who did not return to their pre-activation employer and instead collected UCX, 40 percent listed only voluntary reasons, 34 percent listed both voluntary and involuntary reasons, and 26 percent listed only involuntary reasons.

Implications

Our analyses suggest that the sharp rise in the UCX caseload is not evidence of a substantial weakening of the civilian labor market for recent veterans. Instead, our analyses suggest that the increase in the UCX caseload is due to changes in the population of veterans eligible for UCX. This includes a sharp increase in the number of veterans who qualify on the basis of their reserve service and a sharp increase in the length of reserve deployments.

Nevertheless, the sharp increase in the UCX caseload might suggest a rethinking of the UCX program. It appears that many reservists collecting UCX have a USERRA-protected job to which they have a statutory right to return. Policymakers might consider administrative steps to ensure that reservists claiming UCX understand that a USERRA-protected job will usually make them ineligible for UCX and to ensure that state UI program employees deter-

[2] The availability of a USERRA-protected job does not automatically disqualify reservists from receiving UCX. State workforce agencies make eligibility determinations based on a variety of other considerations, some of which vary from state to state. However, availability of suitable employment is a major consideration.

mine whether claimants have access to a USERRA-protected job before granting UCX benefits.[3] Discussions with DoL ETA staff suggest that official guidance on this issue was ambiguous through mid-2007 and that additional guidance to reservists and to state UI offices and employees might be useful.[4]

Alternatively, policymakers might decide that UCX is an appropriate vehicle for delivering income support to reservists who are recuperating from a stressful deployment and possibly considering alternative civilian employment.[5] It should be noted that any such UCX-funded leave would be in addition to paid leave accumulated while serving on active duty and that USERRA already allows reservists up to three months to return to their pre-activation job (though USERRA itself does not provide any income during that period).

---

[3] We use "policymakers" generically here to refer to federal and state agencies with authority to enforce and interpret UCX and USERRA statutes and regulations.

[4] The ambiguity arises from a disparity between the language contained in the UCX Handbook (DoL ETA, 1994, Chapter IV, 7.a) and that in the Unemployment Insurance Program Letter (UIPL) No. 27-06 (DoL ETA, 2006b). The guidance contained in UIPL No. 27-06, issued in August 2006, is correct. DoL is revising the UCX Handbook to conform to the guidance contained in UIPL 27-06.

[5] Such a change would require a legislative change to the UCX statute, which, as written (20 U.S.C.F.R., Part 614), requires individuals to be able and available to work. Most states also require that individuals be actively seeking work.

Acknowledgments

This research would not have been possible without the assistance of dedicated staff within the Office of the Secretary of Defense (OSD), the individual military services, the Department of Labor, and RAND. John Winkler and James Scott (OSD–Reserve Affairs) and Saul Pleeter (OSD–Military Personnel Policy) provided invaluable guidance throughout the course of the project. We are grateful to Scott Seggerman, Barbara Balison, Darlena Ridler, Christi Phillips (Defense Manpower Data Center [DMDC]), and Major Steven Tobler (Medical Surveillance Activity, U.S. Army) for assisting us with obtaining and interpreting military personnel data. Brian Lappin (DMDC) worked with us to develop questions for the Status of Forces Survey on pre- and post-activation civilian labor market experiences. Peggy Whitney and Glenda Malone (U.S. Army), LT Bradley Lewis and LT Rosie Goscinksi (U.S. Navy), Gail Weber (U.S. Air Force), and Christopher Smith (U.S. Marine Corps) coordinated the retrieval and transmission of UCX records to RAND. Keith Ribnick (DoL ETA) patiently answered our questions about the UCX program and provided us with aggregate tabulations. At RAND, Leah Barnes Calderone, Carl Matthies, and Bogdan Savych provided superior research assistance and Debbie Wesley, Bob Reddick, and Craig Martin provided excellent programming support.

This research also benefited from the thoughtful comments of David Chu (Under Secretary of Defense for Personnel and Readiness), Thomas Hall (Assistant Secretary of Defense for Reserve Affairs), Charles "Chick" Ciccolella (Assistant Secretary of Labor for Veterans' Employment and Training), Leslye Arsht (Deputy Under Secretary of Defense for Military Community Family Policy), V. Penrod (Director of Military Compensation, OSD–Military Personnel Policy), Jane Burke (Principal Director, OSD–Military Community and Family Policy Office), Ruth Samardick (DoL ETA), Ron Horn (OSD–Military Community and Family Policy Office), Tim Elig (Division Chief, Human Resources Strategic Assessment Program, DMDC), Betty Castillo (DoL ETA), Stephanie Garcia (DoL ETA), James Hosek (Director of the Forces and Resource Policy Program within the RAND National Defense Research Institute), James Dertouzos (RAND), and Sebastian Negrusa (RAND).

Abbreviations

ADPF	Active Duty Pay File
CPS	Current Population Survey
DMDC	Defense Manpower Data Center
DoD	U.S. Department of Defense
DoL	U.S. Department of Labor
DVA	U.S. Department of Veterans Affairs
ESGR	National Committee for Employer Support of the Guard and Reserve
ETA	Employment and Training Administration
FCCC	Federal Claims Control Center
GWOT-CF	Global War on Terror Contingency File
MEF	Master Earnings File
OSD	Office of the Secretary of Defense
PDHA	Post-Deployment Health Assessment
PTSD	post-traumatic stress disorder
RPF	Reserve Pay File
SOFS-R	Status of Forces Survey of Reserve Component Members
UC	Unemployment Compensation
UCX	Unemployment Compensation for Ex-Servicemembers
UI	Unemployment Insurance
UIPL	Unemployment Insurance Program Letter
USERRA	Uniformed Services Employment and Reemployment Rights Act
WEX	Work Experience File

CHAPTER ONE

Introduction

The U.S. Department of Defense's (DoD's) Unemployment Compensation for Ex-Servicemembers (UCX) program provides income assistance to unemployed veterans as they search for work. The primary difference between UCX and the civilian Unemployment Insurance (UI) program, after which it is modeled, is in the determination of initial eligibility. To be eligible for UI, workers must have been involuntarily separated from their most recent job. In contrast, honorably discharged active-component personnel and reserve-component personnel completing a period of active-duty service of 90 or more days are eligible to receive UCX benefits regardless of whether their separation was voluntary in nature. UCX eligibility, benefit levels, and duration are otherwise determined by regulations governing the civilian UI program. The federal government dictates certain requirements for UI, but specific program parameters vary significantly from state to state. Broadly speaking, however, UCX recipients must be able and available to work and must be actively seeking work.[1] UCX payments are typically available for as many as 26 weeks and depend on pre-unemployment earnings, number of dependents, and other factors. In 2007, maximum weekly UCX benefits ranged from $235 in Alabama to as much as $862 in Massachusetts (DoL ETA, 2008). UCX benefits are terminated as soon as the individual is reemployed.[2]

Between 2002 and 2004, the UCX caseload, by which we mean the number of individuals receiving UCX benefits, increased by 72 percent (see Figure 1.1). In the first quarter of 2002, an average of 16,243 veterans were receiving UCX benefits in a given week.[3] By the first quarter of 2004, the caseload had increased to 27,986. The average weekly caseload peaked at 28,967 in the first quarter of 2006 and stood at 23,335 in the fourth quarter of 2007.[4]

This sharp increase in the UCX caseload occurred at roughly the same time that official Bureau of Labor Statistics tabulations from the Current Population Survey (CPS) indicated that the unemployment rate of young veterans ages 20–24 was also increasing (from 11.0 per-

---

[1] Federal eligibility requirements are specified in 20 U.S.C.F.R., Part 614. This federal statute requires individuals to be able and available to work. Most states also require that individuals be actively seeking work. It is also typically the case that a refusal of suitable employment renders an individual ineligible to receive UCX benefits, but this determination is adjudicated under state law on a case-by-case basis.

[2] For more information on the civilian UI program and the UCX program, see Appendix A of this document. See also 20 CFR, Part 614, and Stone, Greenstein, and Coven (2007).

[3] The total number of individuals receiving UCX in a given quarter is not computable with these data.

[4] UCX caseloads were much higher in the early 1990s. In the first quarter of 1994, for example, the average weekly UCX caseload was 44,537. The average weekly UCX caseload reached a minimum of 11,805 in the second quarter of 2000. The relatively high caseload of the early 1990s most likely reflects the sharp increase in active-component forces leaving the military as part of the post–Cold War drawdown of active-component forces that occurred during that period.

Figure 1.1
Average Weekly UCX Caseload, by Quarter

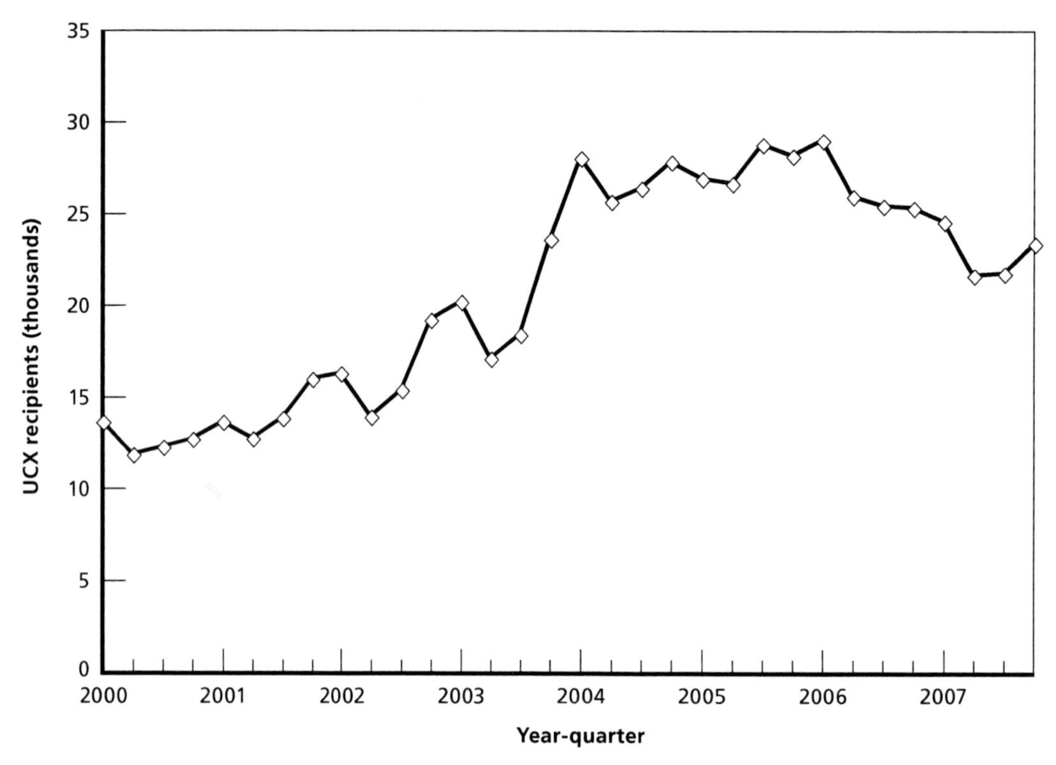

SOURCE: DoL ETA.
RAND *TR588-1.1*

cent in 2003 to 15.6 percent in 2005). The increase in unemployment among veteran youth was particularly worrisome since, between those same years, the overall youth unemployment rate declined.[5]

The analyses of Savych, Klerman, and Loughran (2008) question the practical significance of this increase in veteran youth unemployment between 2003 and 2005, as does the return of the unemployment rate to levels consistent with long-run historical norms in 2006.[6] But the sharp and sustained increase in the UCX caseload over this same period of time remains a cause for concern; specifically, the increase in the UCX caseload could indicate that veterans of the wars in Iraq and Afghanistan are having difficulty transitioning to the civilian labor market and that DoD, the U.S. Department of Labor (DoL), and the U.S. Department of Veterans Affairs (DVA) need to reconsider how they deliver transition assistance to recently demobilized veterans.

---

[5] For example, on February 2, 2006, Senator Larry E. Craig, at hearings of the Senate Veterans Affairs Committee entitled "Jobs for Veterans Act Three Years Later: Are Vets' Employment Programs Working for Veterans?" remarked,

> Ensuring a smooth transition for those young service members who bravely fought for freedom should be a national priority. Yet as the chart behind me shows, these unemployment rates among young veterans have risen dramatically since the war on terror began and now are approaching double the unemployment rate of nonveterans in the same age group. I must add, in a strong economy, as is true in most areas of our country, these figures just don't fit. (U.S. Senate, 2006.)

[6] They conclude that random sampling variation likely accounts for the spike in veteran youth unemployment rate measured in the CPS. They also note that a similar spike in veteran youth unemployment is not evident in the Census Bureau's American Community Surveys.

This report examines the reasons why the UCX caseload has risen and considers the implications of those findings for the UCX program. The next chapter documents aggregate trends in the UCX caseload and numbers of servicemembers eligible to receive UCX, explains how we collected data on individual UCX claims, and shows how the UCX caseload changed over time within specific military components. Chapter Three decomposes the increase in the UCX caseload into two main sources: (1) increases in eligibility and (2) increases in the UCX claim rate. Chapter Four presents evidence on the relationship between deployment duration, post-deployment health, and the UCX claim rate. Chapter Five then examines survey data to further characterize the pre- and post-activation employment experiences of UCX recipients. The final chapter discusses the implications of our empirical findings for veteran unemployment and the structure of the UCX program.

CHAPTER TWO

Trends in the UCX Caseload

In this chapter, we first present time-series plots of the aggregate UCX caseload and numbers of active- and reserve-component members eligible to receive UCX. Understanding these changes requires information on the characteristics of those receiving UCX. The second section of this chapter describes how we built an individual-level database of those receiving UCX. The final section of the chapter uses those individual-level data to show that nearly 60 percent of the increase in the UCX caseload is attributable to increases in the numbers of Army Reserve and National Guard personnel receiving UCX.

Trends in the Aggregate UCX Caseload and Eligible Population

States report weekly claims for each of their unemployment programs, including UCX, to DoL's Employment and Training Administration (ETA). Using these state reports, ETA publishes on its Web site national counts of weekly UCX claims. ETA also compiles data on aggregate UCX claim values and claim duration from these state reports. ETA staff provided us with these later time series.[1] Data on total UCX claims and claim duration are not available by service or component; the aggregate claim value data are available by service but not by component.

Figure 2.1 plots the average weekly UCX caseload by quarter on the left-hand axis and the total value of UCX payments by quarter on the right-hand axis. As shown before in Figure 1.1, the UCX caseload increased by more than 70 percent between 2002 and 2004. The total value of UCX payments, which are funded by the individual military services, also increased sharply over that same period. Total UCX payments were $265 million in 2002 and $497 million in 2004.

Note also that there is a strong seasonal component to the UCX time series. The UCX caseload is generally highest in the first and fourth quarters. These strong seasonal patterns partially obscure the timing of longer-term changes in the UCX caseload. Therefore, in the balance of this section, we present seasonally adjusted UCX counts. When we turn to individual-level data, we compare results in the same calendar quarter. Doing so also eliminates seasonal effects.

One might think that changes in the UCX caseload would be due to changes in the state of the economy. If so, we would expect the UCX caseload to move with the civilian UI case-

[1] Weekly claims reports can be found on the DoL ETA's "UI Weekly Claims" (DoL ETA, no date). We are grateful to Keith Ribnick, UI program specialist at the ETA, for providing us with the UCX claim value and duration data.

Figure 2.1
Average Weekly UCX Caseload and Value of UCX Payments, by Quarter

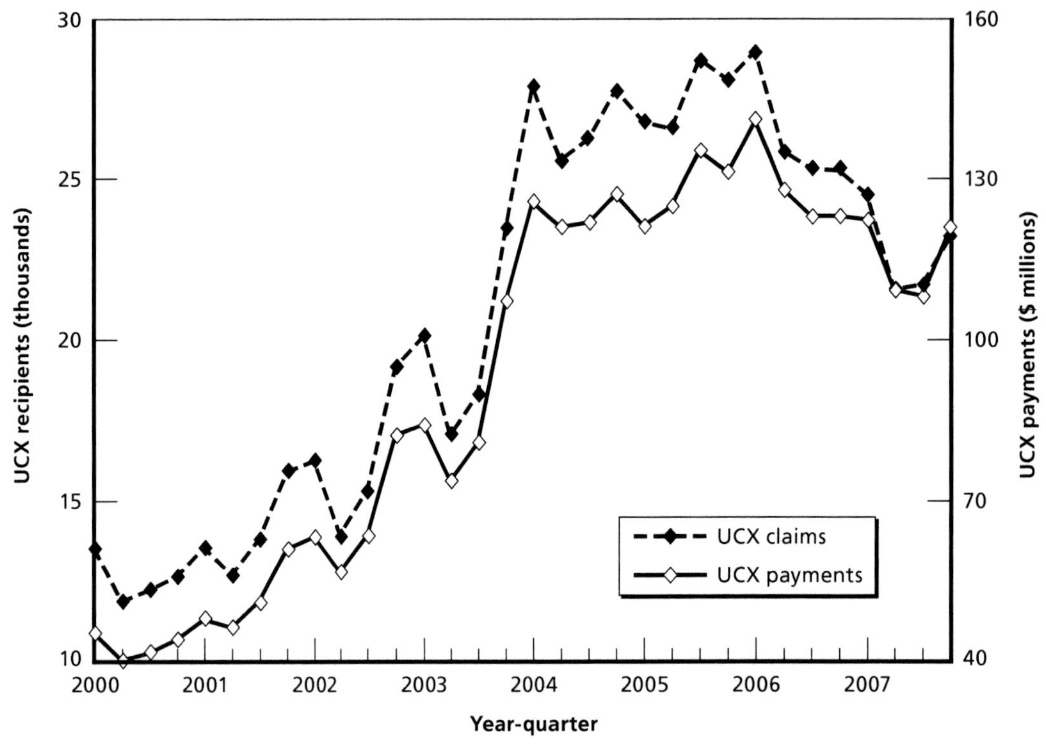

SOURCE: DoL ETA.
RAND *TR588-2.1*

load. To explore the importance of such a relation between the UCX caseload and the state of the economy, Figure 2.2 plots seasonally adjusted aggregate caseloads for regular UI and UCX.[2]

The civilian UI caseload increased sharply between 2000 and 2001, as the U.S. economy entered recession, and remained at an elevated level through mid 2003. The UI caseload fell steadily through early 2006 and then began to trend upward once again through 2007.

Movements in the UCX caseload are quite different. The increase in the UCX caseload did not begin until early 2001. The UCX caseload increased through 2001, fell somewhat in early 2002, and then began to rise again. The sharpest increase in the UCX caseload was between late 2002 and early 2004, a period in which the civilian UI caseload was holding steady and then beginning to decline. Thus, a different set of influences appears to be driving changes in the UCX caseload.

The total UCX caseload consists of both veterans who have just begun to receive UCX benefits and veterans who have been receiving benefits for as many as 26 weeks. Thus, the UCX caseload at any point in time is a function of both the rate at which individuals enter the UCX program and the number of weeks over which they receive benefits. If average claim duration increases, the caseload can continue to rise even if the rate of entry declines.

---

[2] The two series are normalized to an index value of one in the first quarter of 2000. Specifically, the seasonally adjusted series is the residual from a regression of the log of the caseload on four seasonal dummies and normalized to 2001Q1 = 1.0.

Figure 2.2
Seasonally Adjusted Average Weekly UI and UCX Caseload, by Quarter

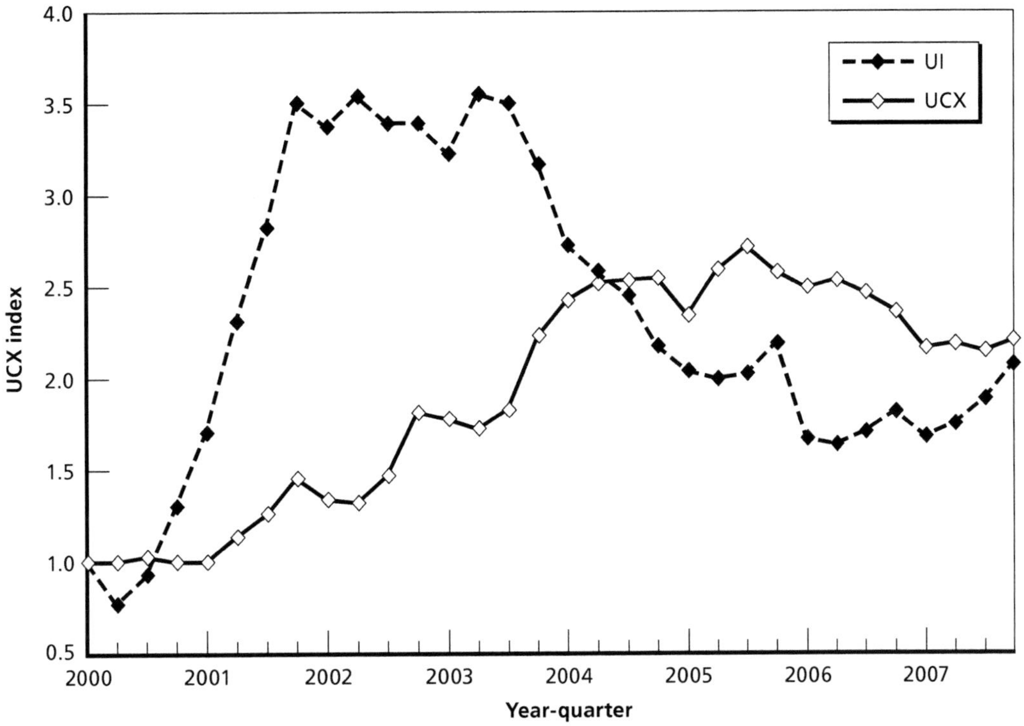

SOURCE: DoL ETA.
NOTE: Index = 1 in 1st Quarter, 2001.
RAND *TR588-2.2*

According to DoL statistics, the average number of weeks of UCX benefits received by UCX recipients increased from 17.6 in 2002 to 19.2 in 2007, and the fraction of UCX recipients who exhausted their benefits increased from 42 to 45 percent.[3] These statistics suggest that the exit rate has declined somewhat over this time period, further fueling the increase in the aggregate caseload.

Initial UCX claims are likely to serve as a more sensitive indicator of the timing of changes in the demand for UCX (claims are not necessarily paid), since demand is a function only of the current period entry rates. In contrast, the total UCX caseload is a function of both earlier entry rates and the rate at which individuals exit the UCX program. Figure 2.3 plots seasonally adjusted initial UCX and UI claims. It shows that initial UCX claims rose between the first and fourth quarters of 2002, declined somewhat, and then rose again between the second quarter of 2003 and the third quarter of 2004.

One obvious hypothesis for the increase in the UCX caseload is an increase in the number of veterans eligible to receive UCX. Even if the fraction of eligible servicemembers who receive UCX is constant, if the number of eligible servicemembers increased, the UCX caseload would continue to increase. Recently separated (honorably discharged) active-component personnel and deactivated reservists who completed at least 90 days of active-duty service are eligible to receive UCX providing they meet other UI requirements. We derive counts of active-duty

[3] Authors' computations based on aggregate UCX data provided by ETA.

Figure 2.3
Seasonally Adjusted Average Weekly Initial UI and UCX Claims, by Quarter

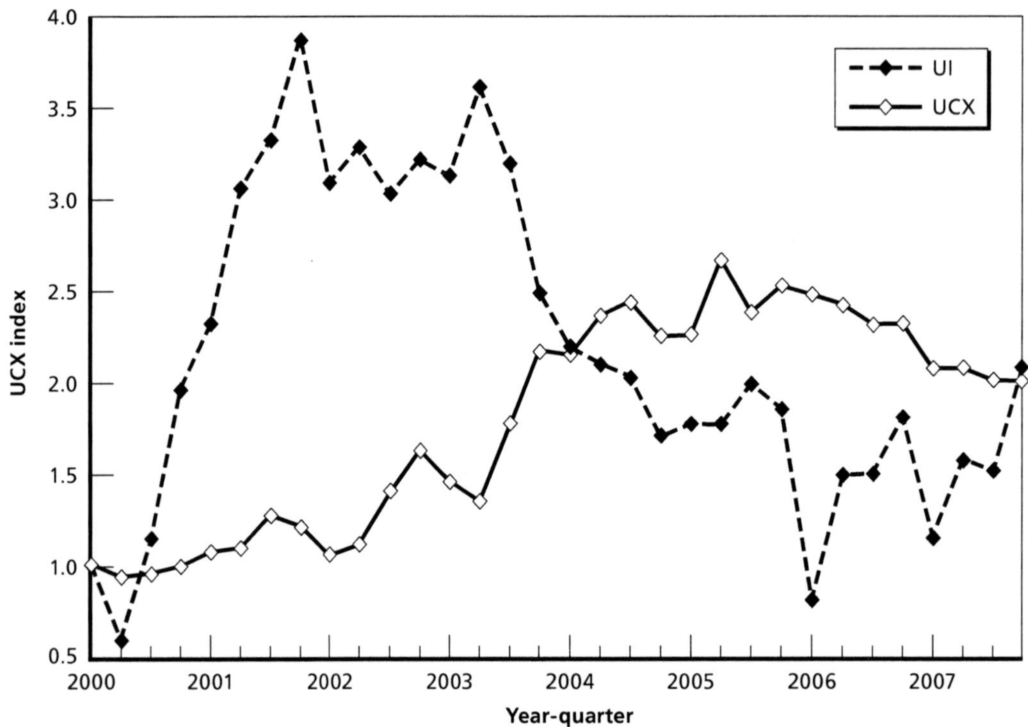

SOURCE: DoL ETA.
NOTE: Index = 1 in 1st Quarter, 2001.
RAND *TR588-2.3*

separations from the Work Experience File (WEX)[4] and counts of reserve deactivations (for activations of at least 90 days) from the Active Duty and Reserve Pay Files (ADPF and RPF, respectively).[5] Both counts exclude Coast Guard veterans.[6]

An alternative approach to generating counts of deactivated reservists would be to use the Global War on Terror Contingency File (GWOT-CF) maintained by the Defense Manpower

---

[4] The WEX is generated from DMDC's Active Duty Military Personnel Master File and Reserve Component Common Personnel Data System File and contains at least one record for every individual serving in the active or reserve components on or after September 30, 1990. The file contains dates of separation for each active-component member. These separations are not necessarily permanent.

[5] The pay files, maintained by DMDC, record all the military pays received by reservists in a given month. We estimate days of active-duty service in a given month from these data by dividing basic pay received by the value of basic pay a particular reservist would receive for one day of active-duty service. Activation spells are then defined by finding blocks of full days of active-duty service. Because the pay data record monthly rather than daily pay totals, we must make a number of assumptions in defining activation spells:

1. A month with 30 active-duty days is assumed to be a contiguous block.
2. A month of less than 30 days that is next to a month with 30 active-duty days is assumed to be one contiguous block (e.g., 15 days in June followed by 30 days in July is assumed to be 45 contiguous days).
3. Any month with more than 25 active-duty days is assumed to be a full 30 days of active-duty (i.e., if active-duty days > 25, then active-duty days = 30).
4. Any month of 20 or more active-duty days that is bounded on both sides by months with 30 active-duty days is considered to be 30 active-duty days.

[6] Less than 1 percent of UCX payments are made to the Coast Guard.

Data Center (DMDC). The GWOT-CF is intended to include a record for every activation or deployment since September 11, 2001, in support of the Global War on Terror and its specific contingencies (Operation Noble Eagle, Operation Enduring Freedom, and Operation Iraqi Freedom). Using the pay file, however, we can capture a larger number of relevant reserve activation spells. For example, the pay files would capture a reservist activated to backfill a military position on a U.S. domestic base left vacant by a deployed active-component servicemember. This type of activation is less likely to be included in GWOT-CF.

As can be seen in Figure 2.4, active-component separations have been fairly stable since 2000, averaging about 47,500 per quarter. Reserve-component deactivations, on the other hand, increased sharply following September 11, 2001, as reservists were deployed to assist with homeland security and to support operations in Afghanistan. And, with the beginning of Operation Iraqi Freedom, reserve deactivations increased to unprecedented levels. Between 2004 and 2006, reserve deactivations (for activations lasting 90 or more days) averaged more than 68,000 per quarter.

These descriptive statistics suggest that the reserve components likely account for a significant fraction of the overall increase in the UCX caseload. We provide further evidence for this conjecture in the third section of this chapter. First, though, we explain how we built the individual-level database on UCX claims that we use to conduct those analyses.

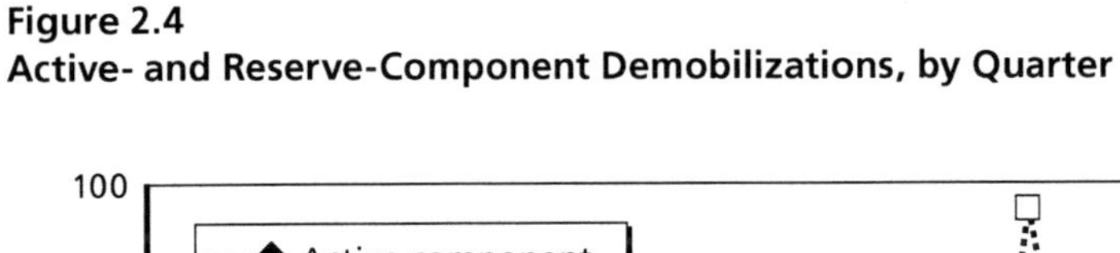

Figure 2.4
Active- and Reserve-Component Demobilizations, by Quarter

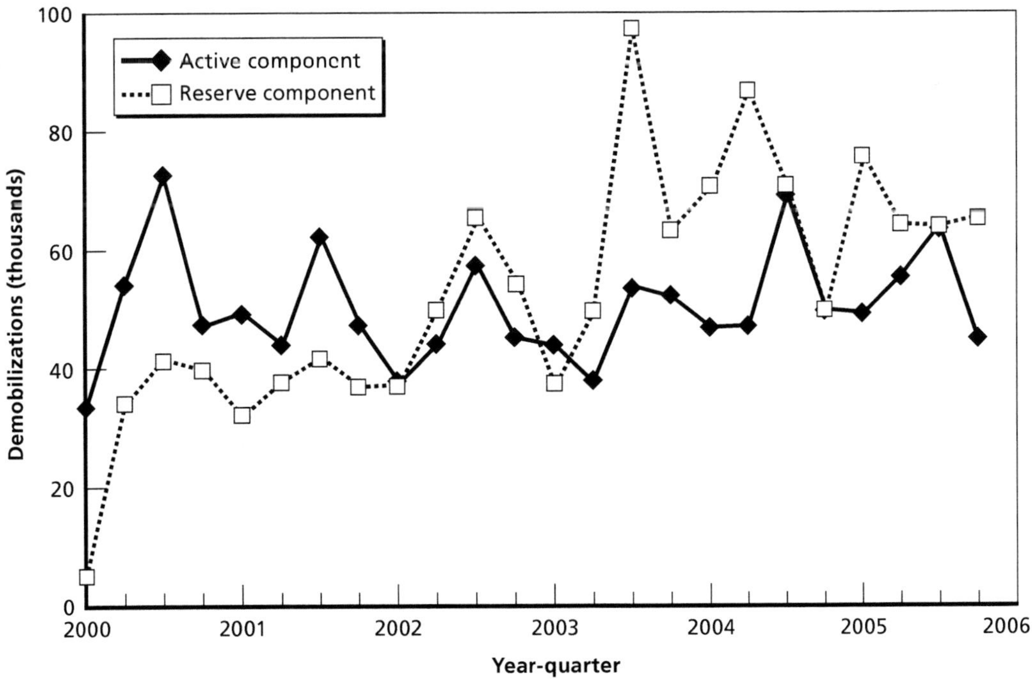

SOURCES: WEX, ADPF, RPF.
NOTE: Counts exclude the Coast Guard.
RAND *TR588-2.4*

Constructing the UCX Database

A significant feature of this research is the use of individual-level data on UCX claims. These individual-level data allow us to estimate how the UCX caseload changed over time by service and component and how other individual-level characteristics influence the propensity to receive UCX benefits. State workforce agencies maintain records of payments made to individuals under UI, UCX, and Unemployment Compensation for Federal Employees (UCFE). Those individual-level data, however, are not compiled into a single national file available to the research community. Some state workforce agencies do share those data with researchers, but this project required a nationally representative sample of UCX recipients, and so obtaining data directly from the small number of state agencies willing to participate would not have been adequate.[7]

Instead, RAND obtained lists of UCX recipients from the individual military services. These lists are generated by each state and provided to the services on a quarterly basis. The individual services use these quarterly reports to verify eligibility of veterans for UCX. The lists also serve as an invoice of sorts, since the individual services must compensate each state for any UCX payments they issue.[8] The data fields on these quarterly reports vary by state but, at a minimum, include the names and Social Security numbers of each individual who received a UCX payment during that quarter and the amount of that payment. In some cases, the quarterly reports also provided information on when the individual first received a UCX payment under this claim.

Each service maintains these quarterly reports for some period of time in hard-copy form. RAND obtained all available quarterly reports from each of the services (excluding the Coast Guard), catalogued those reports, and then made a decision about which quarters of data to have key-punched into electronic form.[9]

Considerations of coverage and cost guided our decision to key-punch two quarters of reports: the second quarter of 2002 for the Army and Air Force and the second quarter of 2005 for all four services.[10] These two quarters span the period over which the aggregate UCX caseload increased and had the most complete geographic coverage in each quarter. Using the same calendar quarter controls for the seasonality in UCX claims observed in Figure 2.1. Unfortunately, the Navy and Marine Corps did not maintain sufficiently complete records for

[7] Another centralized source of information on UCX claimants is the Federal Claims Control Center (FCCC). Each state workforce agency uses the FCCC to verify whether a veteran is in fact a veteran and eligible to receive UCX benefits. Thus, the FCCC maintains data on veterans who file an initial claim for (but do not necessarily receive) UCX benefits. We were unable to secure an agreement with DoL to obtain these data for this project. Even if we had, FCCC maintains only nine quarters of data, which would have necessitated our obtaining data for earlier periods from the individual services.

[8] If it is determined that a UCX applicant has intentionally provided fraudulent information or documentation to receive UCX benefits, states will require repayment of UCX benefits that were paid in error. In other instances, it is up to the state to determine if repayment will be required on a UCX overpayment; however, even if the overpayment is waived, the state is still responsible for repaying the appropriate federal account for benefits that were paid in error.

[9] We attempted to scan the paper records and convert them to a machine-readable format using Optical Character Recognition. However, this process was not sufficiently accurate. Instead, we engaged a contractor to manually key-punch the data contained on the paper records. The contractor entered the data, including Social Security numbers. The contractor then sent that file to DMDC, which converted the Social Security number to a scrambled Social Security number that DMDC provided to RAND to facilitate linking data files. Our analysis files do not contain actual Social Security numbers.

[10] As an initial test, we also key-punched data for the fourth quarter of 2005 for the Army. We do not employ those data in the analyses reported here.

any quarter of 2002. Our completed analysis file contains data from 50 states, the District of Columbia, Puerto Rico, and the U.S. Virgin Islands.[11]

In order to obtain additional information about these UCX recipients, we merged the individual-level UCX records to military personnel records contained in the WEX, GWOT-CF, ADPF, RPF, and the Post-Deployment Health Assessment (PDHA).[12] Service was determined from the UCX records themselves, but applicable component was determined by the WEX and pay files. We were unable to determine component for about 10 percent of veterans in 2002 and for about 5 percent of veterans in 2005.[13] These individuals were dropped from our analysis sample.

The Change in UCX Claims Between 2002 and 2005

Before turning to the individual-level UCX data, we estimate changes in the UCX caseload between the second quarters of 2002 and 2005 by service, employing the aggregate ETA claim value data. We estimate the average weekly UCX caseload by service by dividing average weekly UCX payments for each service by an estimate of average cost per claim. Average cost per claim is estimated by dividing average total weekly UCX payments across all services by the average weekly UCX caseload across all services. These estimates, reported in Table 2.1, suggest that the Army accounts for 70 percent of the increase in the UCX caseload between 2002 and 2005, the Air Force 16 percent, the Navy 10 percent, and the Marine Corps 4 percent.

Table 2.1
Estimated Average Weekly UCX Caseload, by Quarter and Service

	2nd Quarter			
Service	2005	2002	Difference	Percentage Difference
Army	15,712	6,862	8,850	129
Air Force	3,255	1,241	2,014	162
Navy	4,359	3,118	1,241	40
Marine Corps	3,061	2,450	611	25

SOURCE: DoL ETA.

NOTES: Average weekly caseloads are estimated by dividing average weekly UCX payments for each service by an average cost per claim. Average cost per claim is estimated by dividing total UCX payments across all services by the average weekly UCX caseload across all services.

---

[11] About 0.7 percent of the individuals in the UCX data appeared more than once in the same quarter. In about two-thirds of these cases, these individuals appeared twice in the same state in the same quarter. In the other cases, individuals appeared in multiple states in the same quarter. In both cases, we randomly selected just one of the individual's observations.

[12] See Chapter Four for a description of the PDHA.

[13] Component was determined by searching for the most recent active-component separation date in the WEX or the most recent activation end date in the ADPF or RPF. See Chapter Three for a description of how we define an activation episode from the pay files. The individual's component as listed in those files on that date is then assumed to be the relevant component for claiming UCX. We truncate our search at 20 months prior to the end of the quarter. UCX recipients who do not appear in the WEX or pay files within this window are assigned a missing component and so dropped from the sample (about 6.5 percent of the sample). This most likely happens due to misclassification of active-component separations or reserve deactivations as just defined.

Our individual-level UCX data allow us to examine how the Army and Air Force caseloads vary by component (recall that we were unable to obtain complete individual-level data for the Navy and Marine Corps in 2002).

In Table 2.2, we tabulate numbers of UCX recipients by component and quarter using our individual-level UCX data. We see that in 2005 the reserve components account for 52 percent of the Army UCX caseload and 17 percent of the Air Force UCX caseload. Looking over time, our individual-level UCX data suggest that the Army and Air Force UCX caseload increased by 24 and 82 percent, respectively, between 2002 and 2005. The Army and Air Force reserve-component UCX caseloads increased considerably more in both level and percentage terms. The UCX caseload increased by 704, 900, 344, and 272 percent between 2002 and 2005 in the Army Reserve, Army National Guard, Air Force Reserve, and Air National Guard, respectively.

Tables 2.1 and 2.2 together suggest that about 58 percent of the increase in the UCX caseload is attributable to the Army reserve components alone.[14] The Air Force, Navy, and Marine Corps reserve components represent only 5 percent of the UCX caseload in 2005 (Table 2.2) and so cannot contribute significantly to the overall increase in the UCX caseload. Therefore, the active components must account for the balance of the increase in the UCX caseload (about 40 percent). This share is small relative to the contribution of the active components to the total force.

Table 2.2
Number of UCX Recipients, by Quarter and Component

	2nd Quarter			
Component	2005	2002	Difference	Percentage Difference
A. Active Components				
Army	14,331	11,555	2,776	24
Air Force	4,720	2,595	2,125	82
Navy	7,302	—	—	—
Marine Corps	5,028	—	—	—
B. Reserve Components				
Army Reserve	5,135	639	4,496	704
Army National Guard	10,488	1,049	9,439	900
Air Force Reserve	546	123	423	344
Air National Guard	391	105	286	272
Naval Reserve	1,567	—	—	—
Marine Corps Reserve	674	—	—	—
Total	50,182	—	—	—

SOURCE: RAND UCX database.

[14] From Table 2.1, we estimate that the Army accounts for 70 percent of the increase in the UCX caseload. Table 2.2 suggests that the Army reserve components account for 83 percent of the increase in the combined Army active- and reserve-component caseload. So, by these figures, about 58 percent (83 × 70 percent) of the total increase in the UCX caseload is attributable to the Army reserve components.

CHAPTER THREE

The Effect of Eligibility and Claim Rates on the UCX Caseload

In this chapter, we decompose the change in the UCX caseload into a part attributable to the change in the number of veterans eligible to receive UCX and a part attributable to a change in the likelihood that an eligible veteran will claim UCX (the claim rate). The first section describes this decomposition. In the second section, we perform the decomposition, employing individual-level UCX data for the Army and Air Force for the second quarters of 2002 and 2005. We also report on eligibility and claim rates for the Navy and Marine Corps for the second quarter of 2005. Because we do not have Navy and Marine Corps data for 2002, we cannot compute the decomposition for these two services. However, as estimated in Chapter Two, the Navy and Marine Corps account for a small share (14 percent) of the increase in the UCX caseload between 2002 and 2005, and so omitting these services from our analyses is not likely to affect our general conclusions.

Decomposing the UCX Caseload

The number of individuals receiving UCX for the first time in a given period (UCX_I_t) is the product of the number of veterans eligible to receive UCX in that period (E_t) and the fraction of those individuals who claim and receive UCX in that period. We refer to this second parameter as the take-up rate, α_t:

$$UCX\_I_t = E_t \, \alpha_t \tag{3.1}$$

However, we do not have individual-level data on initial UCX claims. Instead, we observe the entire UCX caseload at two points in time, the second quarters of 2002 and 2005. These caseloads can be expressed as

$$UCX_q = \sum_{m=s}^{M} E_{mq} c_{mq} \tag{3.2}$$

where UCX_q is the total UCX caseload in quarter q, E_{mq} is the number of veterans in month m eligible to receive UCX in quarter q (where month m is relative to the end of quarter q), and c_{mq} is the fraction of newly eligible veterans in month m who subsequently receive UCX in any of the three months of quarter q. We refer to this fraction as the claim rate. The first month, s, in which veterans could become eligible to receive UCX in quarter q is at least as far back as 6 months (26 weeks) prior to the first week of the beginning of the quarter. But it could be considerably further back in time, since veterans have some number of months following separation/deactivation to make a claim for UCX. In practice, we set $s = -20$ (i.e., 20 months prior to the last month of the quarter). The last month of eligibility, $m = 0$, is the last month of the quarter.

We define the eligible population in month m as veterans separated from the active components in that month and reservists having completed an activation of 90 or more days in that month. As defined, this population is potentially eligible rather than necessarily actually eligible. Some veterans who meet the requirements just stated may be ineligible to receive UCX for other reasons that we cannot identify in our data. For example, some deactivated reservists might have ready access to suitable employment through a pre-activation employer, which would typically make them ineligible to receive UCX (see Chapter Five for more on this). Nonetheless, for expositional ease, we refer to this population as simply "eligible" rather than "potentially eligible" throughout Chapters Three and Four.

The claim rate as just defined serves as a lower bound on the take-up rate. Suppose, for example, that 100 veterans were separated/deactivated in January 2005 and that 10 of them appear in the UCX data in the second quarter of 2005, implying a claim rate of 10 percent. Some of these 100 eligible veterans might have claimed UCX in the first quarter of 2005 and then exited the UCX program prior to the second quarter of 2005, and some of these eligible veterans might not claim UCX until some date after the second quarter of 2005. Thus, the true take-up rate, which we cannot estimate with our data without making assumptions about exit rates and lags in UCX take-up, will exceed our "claim rate."

Equation 3.2 permits a simple decomposition of the change in the UCX caseload between the second quarters of 2002 and 2005 into a part that is attributable to changes in eligibility and a part that is attributable to changes in claim rates. This decomposition is accomplished by evaluating the change in the UCX caseload in each month m, holding the claim rate constant at the average of the claim rate across the second quarters of 2002 and 2005 in month m, and then evaluating the change in the UCX caseload in each month m, holding eligibility constant at the average level of eligibility across these two quarters in month m. These two quantities sum (approximately) to the actual change in the UCX caseload in month m.[1]

The Effect of Eligibility and Claim Rates on the UCX Caseload

In this section, we first present plots of Army and Air Force eligibility and claim rates relative to the second quarters of 2002 and 2005 and then decompose the total change in these UCX caseloads into parts attributable to changes in eligibility and changes in claim rates.

Changes in UCX Eligibility

In Figures 3.1–3.4, we plot the number of active- and reserve-component veterans potentially eligible to receive UCX benefits by month prior to the end of the second quarters of 2002 and

---

1 For a given month m,

$$dUCX_{mq} \cong \bar{c}_{mq} dE_{mq} + \bar{E}_{mq} dc_{mq}.$$

2005. We generate these tabulations from the WEX for the active components and from the ADPF and RPF for the reserve components.[2]

Figure 3.1 plots the number of eligible active-component veterans by service and month prior to the end of the second quarter of 2005, and Figure 3.2 does the same for the reserve components (month "0" is June 2005 ["*m*" in Equation 3.2], and month "–20" is October 2003). Figures 3.1–3.4 show that eligible veterans are more numerous in the Army active and reserve components than in the other services and that eligibility is highest overall in the Army reserve components.

Figures 3.3 and 3.4 show the change in the eligible population for the Army and Air Force active and reserve components between 2002 and 2005. While the eligible population in the active components does not change appreciably between 2002 and 2005, the eligible population grows considerably in the reserve components, and especially in the Army.

Changes in UCX Claim Rates

Figure 3.5 plots active-component claim rates by service and month prior to the end of the second quarter of 2005. As we would expect, claim rates are highest in the months just pre-

Figure 3.1
Number of Active-Component Veterans Potentially Eligible to Receive UCX in the Second Quarter of 2005, by Months Prior to June 2005 and Service

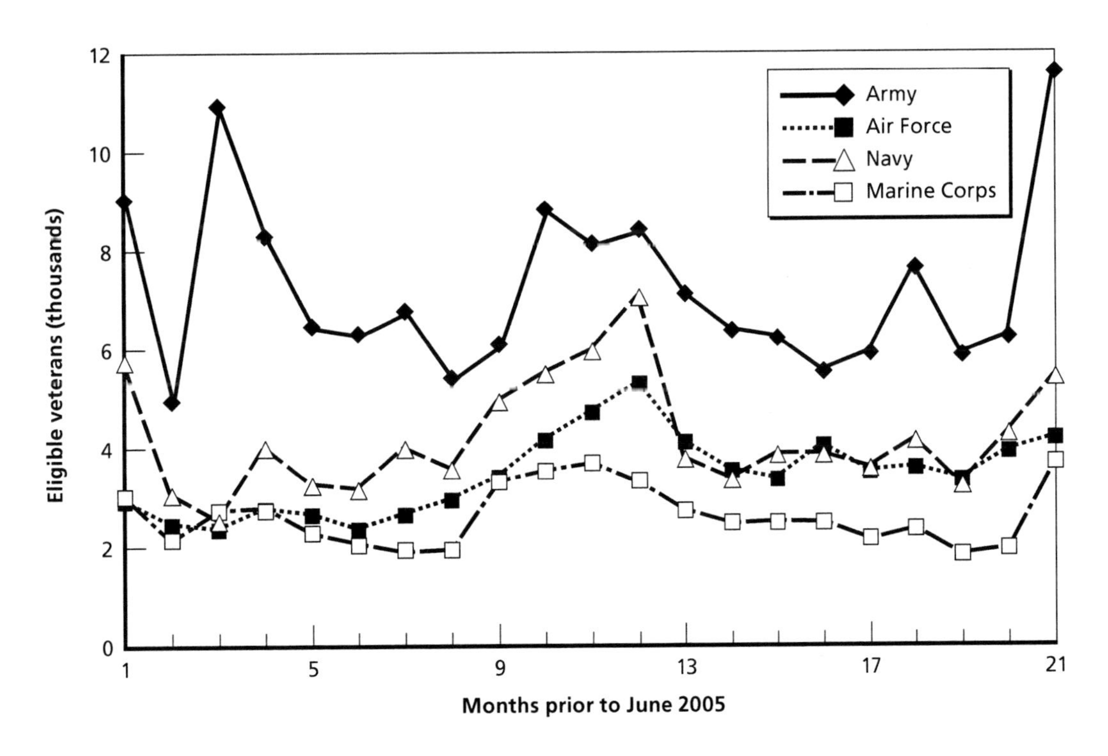

SOURCE: WEX.
RAND *TR588-3.1*

[2] According to DMDC, the Air Force WEX separation data are invalid for June–September 2001. During those months, the Air Force was updating its reporting systems and was unable to transmit valid data to DMDC. DMDC, however, believes that the separation counts reported in the WEX for September 2001 are the sum of the separations for June to September. Therefore, we allocate those separations evenly across those months. We further assume that the claim rate for June to September is equal to the September claim rate.

Figure 3.2
Number of Reserve-Component Veterans Potentially Eligible to Receive UCX in the Second Quarter of 2005, by Months Prior to June 2005 and Service

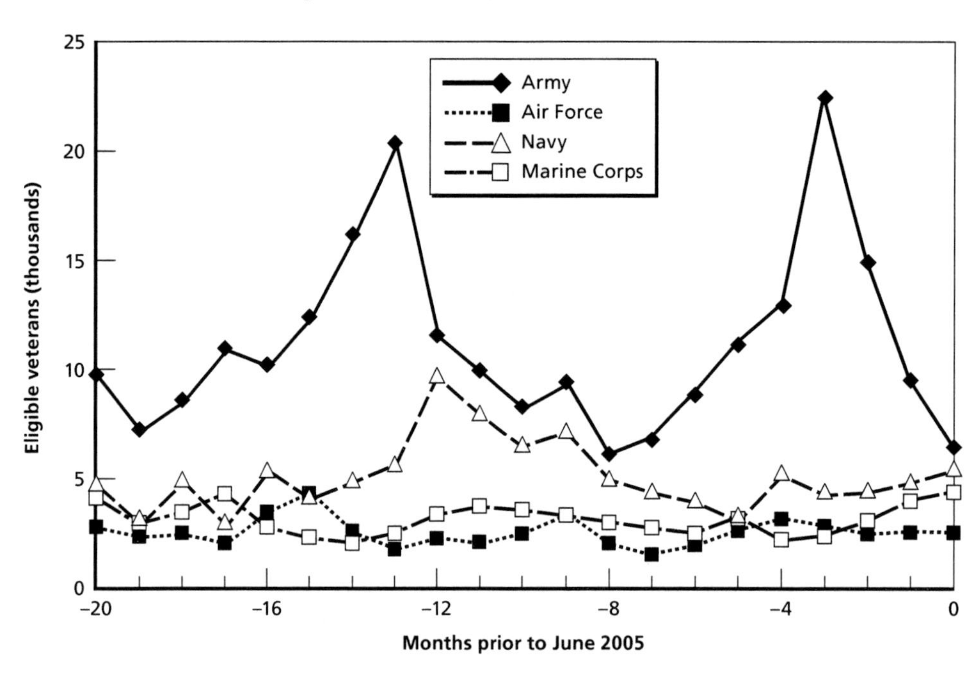

SOURCES: ADPF, RPF.
RAND *TR588-3.2*

Figure 3.3
Number of Army and Air Force Active-Component Veterans Potentially Eligible to Receive UCX in the Second Quarters of 2002 and 2005, by Months Prior to June 2002 and June 2005

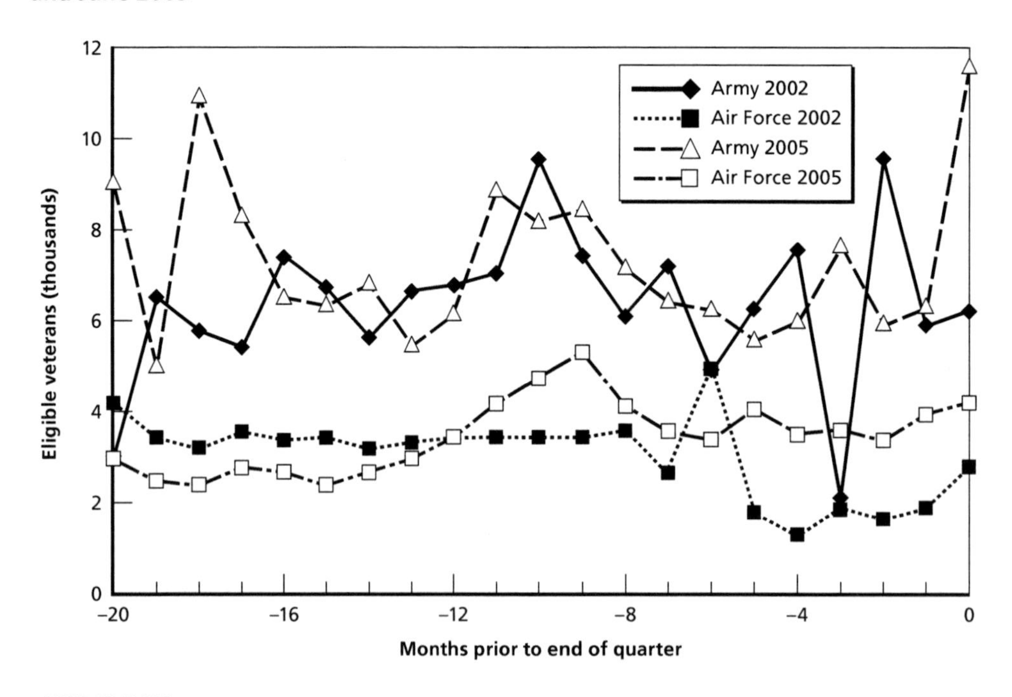

SOURCE: WEX.
RAND *TR588-3.3*

Figure 3.4
Number of Army and Air Force Reserve-Component Veterans Potentially Eligible to Receive UCX in the Second Quarters of 2002 and 2005, by Months Prior to June 2002 and June 2005

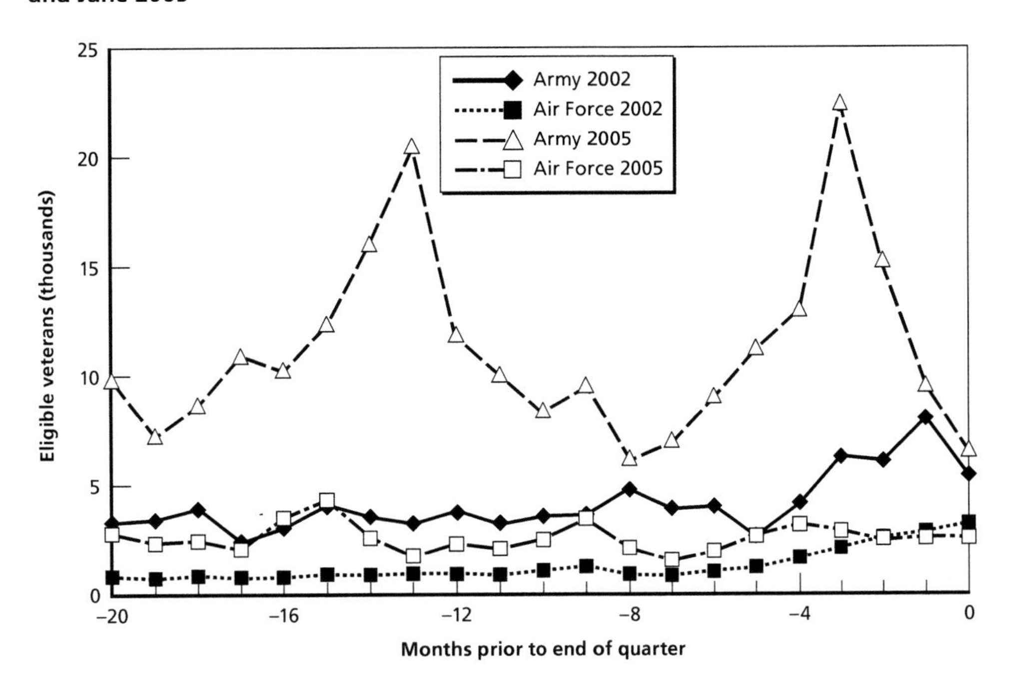

SOURCES: ADPF, RPF.
RAND *TR588-3.4*

Figure 3.5
Active-Component UCX Claim Rates in the Second Quarter of 2005, by Months Prior to June 2005 and Service

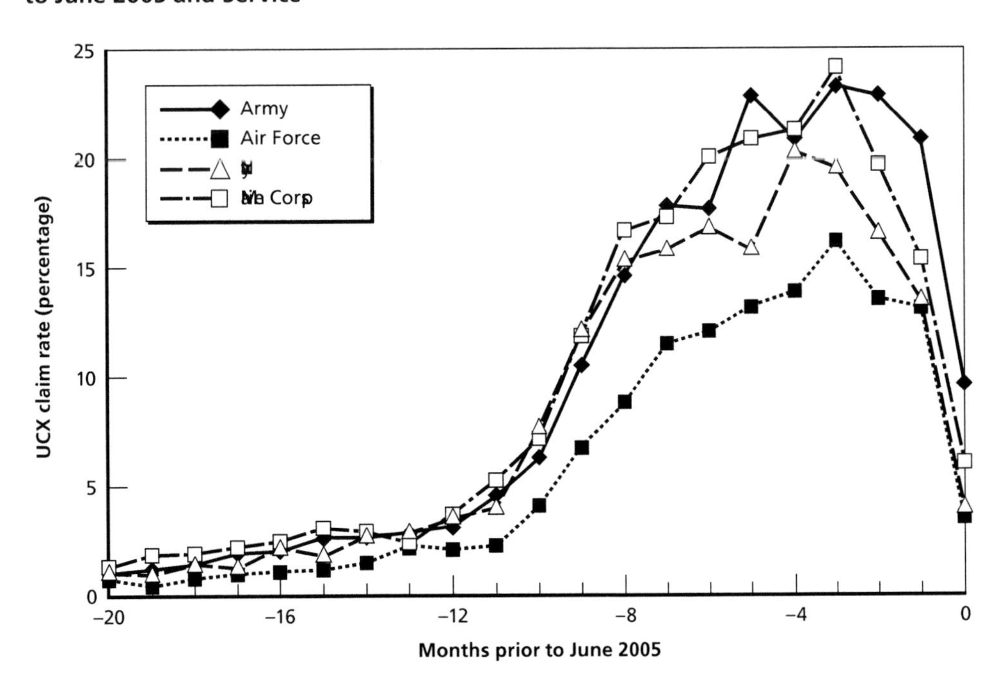

SOURCE: RAND UCX database.
RAND *TR588-3.5*

ceding the beginning of the quarter and approach zero for veterans separated as many as 20 months prior to the end of the quarter. The longer the interval from separation to the quarter of our UCX data, the more likely it is that veterans who claimed UCX will have exited the program prior to that quarter.

There are a number of reasons why veterans who separated so many months prior to the second quarter of 2005 would be receiving UCX in that quarter. First, veterans have as many as 18 months to file a claim for UCX.[3] Second, veterans can receive UCX benefits for up to 26 weeks (i.e., 6 months). Furthermore, under some circumstances, a veteran who is employed soon after separation, but then becomes unemployed shortly thereafter, might still be able to claim UCX. Finally, data errors may lead us to classify a servicemember as having separated at an earlier date than they in fact did.

Figure 3.5 shows that claim rates are highest in the Army and Marine Corps. Their claim rates average 19 and 14 percent, respectively, in the first two quarters of 2005. Claim rates in the Air Force and Navy in those same quarters average 12 and 14 percent, respectively. Thus, a sizable fraction of active-component veterans claimed and received UCX benefits in 2005 (and recall that these figures most likely underestimate the true take-up rate).[4]

Figure 3.6 shows that Army and Air Force active-component claim rates increased somewhat between 2002 and 2005. Averaging over the first two quarters of 2002 and 2005, active-component claim rates increased by 3 and 0.3 percentage points in the Army and Air Force, respectively.

Figure 3.7 plots claim rates in 2005 for the reserve components. Here we see larger differences in claim rates across services. Over the first two quarters of 2005, the Army reserve-component claim rate averages 14 percent. The corresponding claim rates in the Air Force, Navy, and Marine Corps reserve components are 3, 3, and 2 percent, respectively.[5]

Figure 3.8 shows that the Army reserve-component claim rate increased significantly between 2002 and 2005. Averaging over the first two quarters of 2002 and 2005, reserve-component claim rates increased by 11 and 2 percentage points in the Army and Air Force, respectively. These figures represent a five-fold increase in the Army reserve-component UCX claim rate and a four-fold increase in the Air Force reserve-component UCX claim rate.

Decomposition

Clearly, both eligibility and claim rates have changed in ways that would increase the UCX caseload. On average, eligibility increased by 14 percent and 12 percent between 2002 and 2005, and claim rates increased by 0.6 and 2.5 percentage points, in the Army and Air Force active components, respectively (Table 3.1). Our decomposition implies that increased eligibility accounts for 26 and 55 percent of the increase in the active-component UCX caseload

---

[3] Most states have a base period that uses the first four of the last five completed quarters of wages to determine eligibility for benefits. Thus, most servicemembers have approximately 18 months to file an initial UCX claim after separation based on their military wages. After that period, the UCX wages would not be in base period and, therefore, would be irrelevant for computing unemployment benefits. See Appendix A for more about the computation of UCX benefits.

[4] Active-component claim rates are highest among E3–E5s (17–25 percent). Junior officers (O1–O2) have higher claim rates than more senior officers, and officers overall have lower claim rates than enlisted members.

[5] We observe the same pattern in claim rates by pay grade in the reserve components as we observe in the active components (see the previous footnote).

Figure 3.6
Army and Air Force Active-Component UCX Claim Rates in the Second Quarters of 2002 and 2005, by Months Prior to June 2002 and June 2005

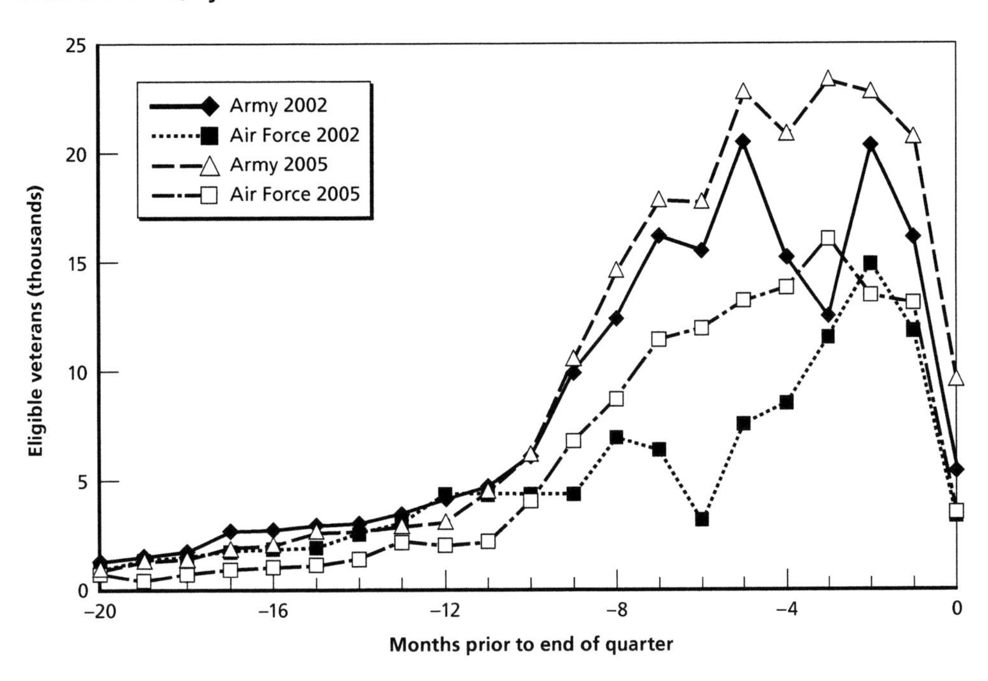

SOURCE: RAND UCX database.
RAND TR588-3.6

Figure 3.7
Reserve-Component UCX Claim Rates in the Second Quarter of 2005, by Months Prior to June 2005 and Service

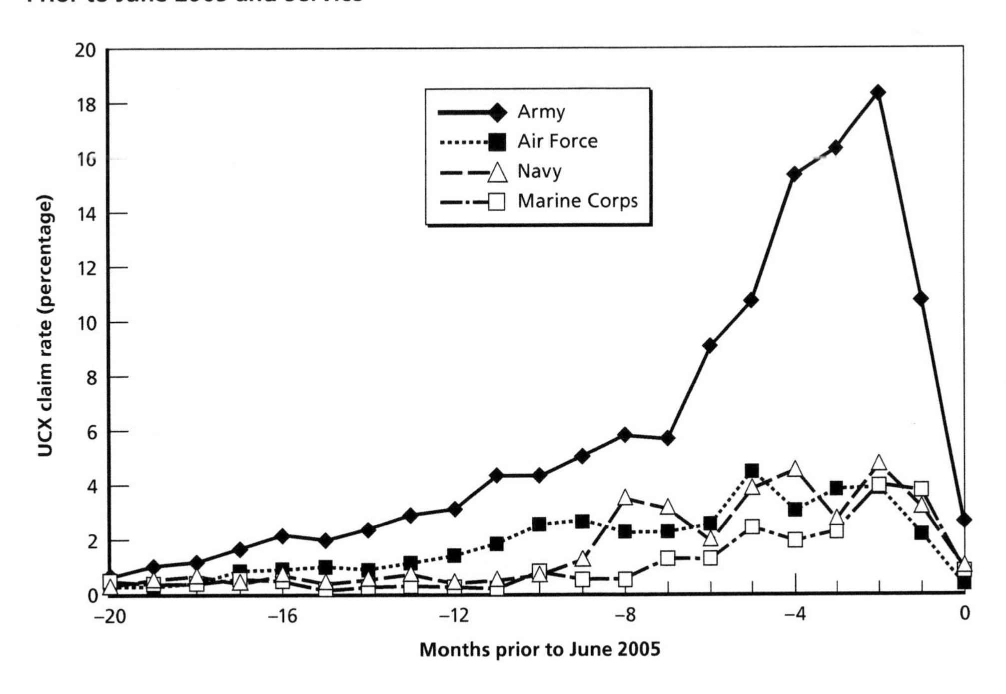

SOURCE: RAND UCX database.
RAND TR588-3.7

Figure 3.8
Army and Air Force Reserve-Component UCX Claim Rates in the Second Quarters of 2002 and 2005, by Months Prior to June 2002 and June 2005

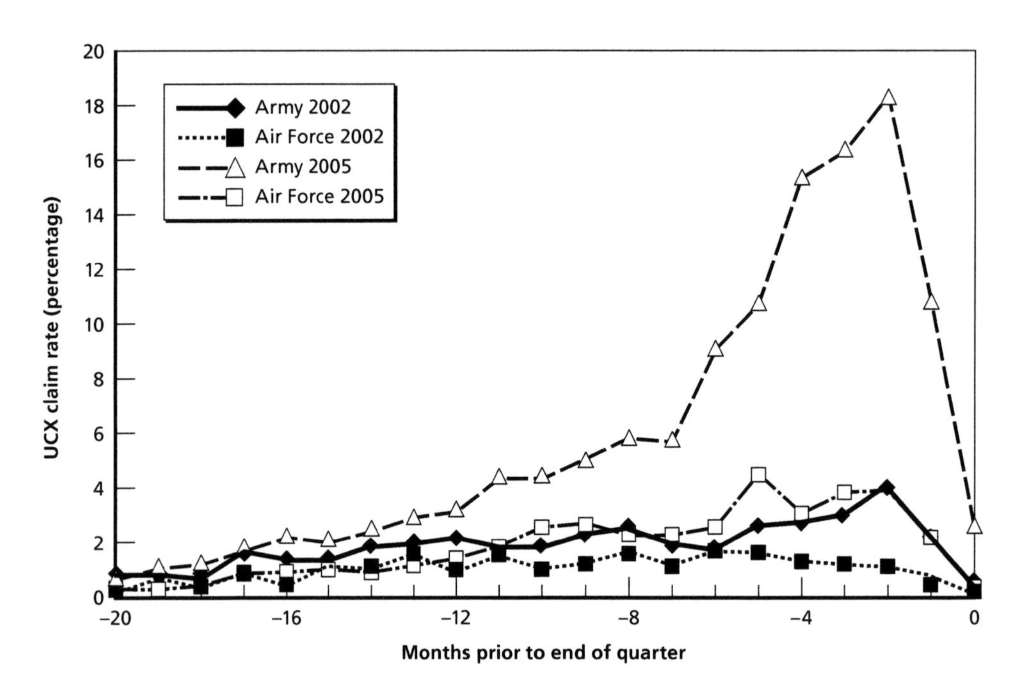

SOURCE: RAND UCX database.
RAND *TR588-3.8*

between 2002 and 2005 in the Army and Air Force, respectively. The increase in claim rates accounts for 74 and 45 percent of these increases in the active-component UCX caseload.

In the reserve components, eligibility increased by 175 and 106 percent and claim rates by 4.8 and 0.9 percentage points in the Army and Air Force, respectively. Our decomposition implies that the increase in eligibility accounts for 46 and 48 percent of the increase in the reserve-component UCX caseload in the Army and Air Force, respectively. The increase in claim rates accounts for 54 and 52 percent of the increase in the reserve-component UCX caseload.

Table 3.1
Percentage of Change in Army and Air Force UCX Caseloads Explained by Change in UCX Eligibility and Claim Rates, 2002–2005

	Eligibility		Claim Rate	
Component	Change (2005–2002)	Percentage Change in UCX Explained	Change (%) (2005–2002)	Percentage Change in UCX Explained
A. Active Components				
Army	19,029	26	0.6	74
Air Force	8,859	55	2.5	45
B. Reserve Components				
Army	148,851	46	4.8	54
Air Force	27,113	48	0.9	52

SOURCE: RAND UCX database.

CHAPTER FOUR

The Effect of Deployment Duration and Post-Deployment Health on UCX Claim Rates

That increases in the UCX caseload have been driven in part by increases in the size of the eligible population is unsurprising. However, that 54 percent of the active and 53 percent of reserve Air Force and Army caseloads are attributable to an increase in the UCX claim rate is worthy of further examination. In this chapter, we explore two reasons why the Army active and reserve UCX claim rates might have risen between 2002 and 2005. We focus on the Army components for two main reasons. First, the Army components account for 70 percent of the increase in the total UCX caseload between 2002 and 2005. Second, claim rates in the Army reserve components were much higher than in the other reserve components in 2005 and increased by the greatest amount between 2002 and 2005 (by 11 percentage points if we look at the first seven months prior to the end of the second quarters of 2002 and 2005). By contrast, UCX claim rates increased by 3, 5, and 5 percentage points in the Army active component, Air Force active component, and Air Force reserve component, respectively.

We propose two broad hypotheses for the increase in the Army UCX claim rate: (1) The civilian labor market for veterans deteriorated, making UCX receipt relatively attractive, and (2) the nature of military service changed, making transitions to the civilian labor market more difficult.[1] We believe that the first hypothesis is unlikely to hold. Between 2002 and 2005, the civilian labor market generally improved; for example, civilian UI claims decreased markedly between these two years (see Figure 2.2). The unemployment rate of veterans ages 20–24 increased between 2003 and 2005 (while the unemployment rate of nonveteran youth declined), but Savych, Klerman, and Loughran (2008) argue that this trend most likely reflects sampling variation rather than real changes in the veteran youth unemployment rate.[2]

The nature of military service, however, has changed markedly over this period of time. For both active- and reserve-component members, the duration and frequency of deployments to combat zones has increased, making military service significantly more demanding and

1 Another possibility, which we do not address in this research, is that the Army population changed along unobservable dimensions correlated with higher UCX claim rates. For example, Army reserve service may have become relatively more attractive to individuals with poor labor market opportunities, and these individuals may be more likely to claim UCX. Although we cannot directly test this hypothesis, we believe that this potential effect is likely to be of second-order importance. The fraction of non-prior-service Army accessions with a regular high school diploma fell from 86 to 84 percent between fiscal years 2002 and 2005. The fraction scoring above the 50th percentile on the Armed Forces Qualifying Test fell from 71 to 69 percent (DoD, 2003 and 2006). However, the changing composition of the Army population is likely to be more of an issue for future research on this topic. The fraction of non-prior-service Army accessions with a regular high school diploma fell from 84 to 70 percent between fiscal years 2005 and 2007 (National Priorities Project, no date)

2 The unemployment rate of veterans 25 and older tracked the civilian unemployment rate between these years.

stressful. Deployed soldiers are separated from family, friends, and a familiar environment and face a significantly higher risk of injury and witnessing others be injured. Most deployed reservists are also separated from their civilian job, which can be stressful in its own right. These additional stresses could make reentering the civilian labor market more challenging and, hence, UCX more attractive. Long deployments might also cause some reservists to reevaluate their civilian careers and consider searching for a new job when they return home. UCX benefits can help support that search process.

In this chapter, we first establish that Army UCX claim rates are positively correlated with length of deployment. We then show that a substantial fraction of the overall increase in the Army reserve-component UCX caseload between 2002 and 2005 is attributable to longer deployments (as we explain below, a similar computation cannot be made for the Army active components). We then establish that post-deployment health worsens with deployment duration and that poor post-deployment health and UCX claim rates are positively correlated. While these later correlations are statistically significant, we argue that they are not large enough in magnitude to explain much of the overall increase in the Army UCX caseload.

Deployment Duration and UCX Claim Rates

We employ the GWOT-CF to compute length of deployment for active-duty members eligible to receive UCX in the second quarter of 2005.[3] For each eligible active-duty veteran, we examine the length of the most recent deployment prior to separation. This computation cannot be made for active-duty members eligible to receive UCX in the second quarter of 2002 since the GWOT-CF begins in September 2001 and the most recent deployment for many eligible active-duty veterans will have begun prior to that date. For Army reservists, we employ the RPF to compute length of deployment, which allows us to examine length of deployment for reservists eligible to receive UCX in either quarter.

Figure 4.1 shows how average UCX claim rates in the first two quarters of 2005 vary with length of deployment. In the Army active components, the UCX claim rate is 13 percent among veterans who were not deployed between September 11, 2001, and their date of separation. UCX claim rates jump to 24 percent for active-component Army veterans who were deployed for one to three months, but do not continue to increase for active-component veterans deployed for longer durations. UCX claim rates increase steadily with deployment duration in the Army reserve components. Claim rates increase from 2, to 7, to 15, to 19 percent for Army reservists activated 4–6, 7–9, 10–15, and 16 or months, respectively.[4]

---

[3] Each record in the GWOT-CF includes the start and end date of each activation or deployment. Generally, deployments are nested within an activation spell. However, there are instances of deployments that occur without a corresponding activation spell or which are not nested within an activation spell. In these cases, we use the union of activation and deployment spells.

[4] This positive correlation between UCX claim rates and deployment duration is independent of pay grade.

Figure 4.1
Army UCX Claim Rates, by Deployment Duration and Component, 2005

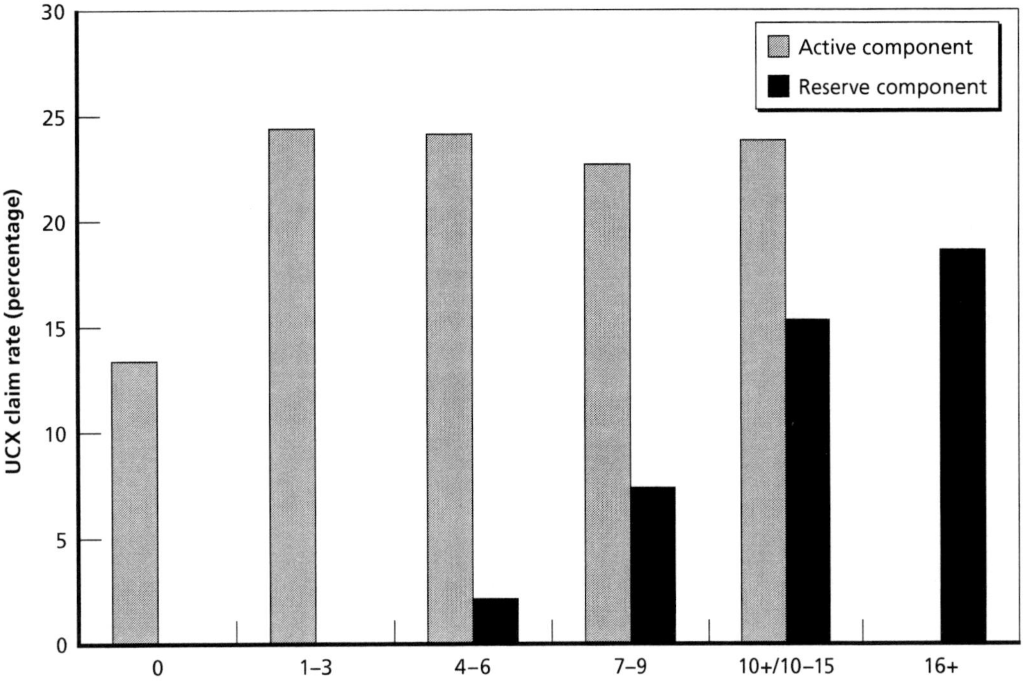

SOURCE: RAND UCX database.
NOTE: UCX claim rates are computed over the first two quarters of 2005.
RAND *TR588-4.1*

Deployment Duration and the Army Reserve Caseload

Deployment duration in the Army reserve components increased sharply between 2002 and 2005; 61 percent of reservists completing 90 or more days of active-duty service in the first two quarters of 2002 were deployed for 4–6 months, and 3 percent were deployed 16 or more months. Far fewer (19 percent) Army reservists were deployed for 4–6 months in 2005, while far more (47 percent) were deployed for 16 or months. Thus, it would appear that part of the increase in UCX claim rates is attributable to longer deployments. Figure 4.2, however, shows that while UCX claim rates increase with deployment duration in both years, UCX claim rates are higher in 2005 for every deployment duration, suggesting that rising deployment duration cannot explain all of the increase in claim rates.

These stylized facts suggest generalizing the decomposition of the UCX caseload in the following way:

$$UCX_q = \sum_{m=s}^{M} \sum_{d=1}^{D} E_{mqd} c_{mqd} \quad (4.1)$$

where eligibility and claim rates are computed by month and deployment duration category (4–6, 7–9, 10–15, and 16 or more months). Employing the same methods used to generate the decomposition reported in Table 3.1, we find that 36 percent of the increase in the Army reserve-component UCX caseload between 2002 and 2005 is attributable to an increase in

Figure 4.2
Army Reserve-Component UCX Claim Rates, by Deployment Duration and Year

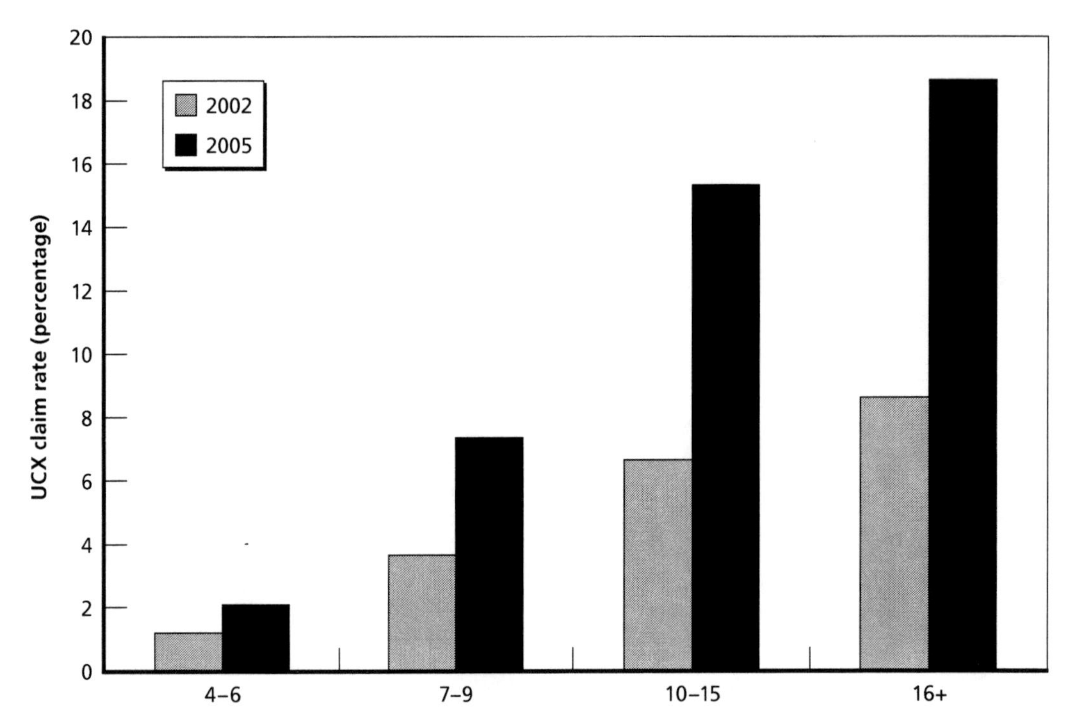

SOURCE: RAND UCX database.
NOTES: UCX claim rates are computed over the first two quarters of 2002 and 2005. Percentages arrayed along the x-axis represent the duration distribution of Army reserve-component activations.
RAND *TR588-4.2*

deployment duration, 46 percent of the increase is due to rising eligibility, and 18 percent to increases in claim rates conditional on duration.[5]

UCX Claim Rates and Post-Deployment Health

With longer deployment comes an elevated risk of injury, and injured veterans could require more time to transition to civilian employment. Although UCX is not intended to provide income support to veterans who are not able and available to work because they are injured, some injured veterans might choose to apply for UCX benefits regardless. In some cases, these veterans might be awaiting a determination of long-term disability status (and disability income

---

[5] To illustrate the decomposition, assume that there are only two deployment lengths, short and long. Define alpha to be the fraction of eligible individuals whose deployments are short. Then,

$$UCX_{mq} = \alpha_{mq} E_{mq} c_{mq} + (1-\alpha_{mq}) E_{mq} c_{mq}$$

and

$$dUCX_{mq} \cong \bar{\alpha}_{mq} \bar{E}_{mq} dc_{mq} + \bar{\alpha}_{mq} \bar{c}_{mq} dE_{mq} + \bar{E}_{mq} \bar{c}_{mq} d\alpha_{mq}$$
$$+ (1-\bar{\alpha}_{mq}) \bar{E}_{mq} dc_{mq} + (1-\bar{\alpha}_{mq}) \bar{c}_{mq} dE_{mq} + \bar{E}_{mq} \bar{c}_{mq} d(1-\alpha_{mq}).$$

support) from DoD and DVA. In other cases, these veterans might be using UCX as a form of short-term disability income while they recuperate from their injuries. In this section, we employ data from the Post-Deployment Health Assessment to show that UCX receipt in the Army components is positively correlated with poor post-deployment health.

PDHA Data

All active and reserve-component servicemembers deployed outside the continental United States to a land-based location with no fixed U.S. medical treatment facility for 30 or more continuous days must complete the Post-Deployment Health Assessment (DD Form 2796) upon redeployment.[6] As stated on DD Form 2796, the principal purpose of the PDHA is "to assess your state of health after deployment outside the United States in support of military operations and to assist military healthcare providers in identifying and providing present and future medical care to you." To this end, the PDHA records information about current physical and mental health, as reported by the servicemember, and documents concerns regarding exposure to environmental toxins, viruses, and the like. The PDHA process includes a face-to-face interview with a health care professional, and the results of that interview are also recorded on the PDHA form along with any referrals for follow-up medical care. While the PDHA process has existed since 1998, it was not fully implemented until 2002–2003.[7]

In order to analyze the correlation between UCX receipt and post-deployment health, we limit our sample to Army active- and reserve-component members who were eligible to receive UCX in the second quarter of 2005. We focus on the Army because PDHA completion rates are highest for this service.[8] We define the eligible active-component population as veterans who separated in the first two quarters of 2005. We define the eligible reserve-component population as veterans who completed an activation of 90 or more days (according to GWOT-CF) in the first two quarters of 2005.[9]

For active-component members, we then search for their most recent deployment, as recorded in GWOT-CF, and match that deployment to a servicemembers' particular PDHA form by the date the servicemember reported that PDHA deployment ended.[10] Using this method, we were able to find PDHA forms for 84 percent of eligible Army active-component servicemembers who had been deployed since September 11, 2001 (56 percent of eligible Army active-component members had no previous deployment and were excluded from the analyses

---

[6] DD Form 2796 is duplicated in Appendix B.

[7] Office of the Chairman of the Joint Chiefs of Staff (2002).

[8] Match rates were 84, 67, 76, and 43 percent in the Army, Air Force, Marine Corps, and Navy active components, respectively. Match rates were 90, 38, 62, and 81 percent in the Army, Air Force, Marine Corps, and Navy reserve components, respectively.

[9] We obtain a higher match rate with the PDHA data when we define activations according to the GWOT-CF as opposed to the pay files and, since in these analyses we are examining UCX receipt only in 2005, we do not need the pay files to define eligibility for the 2002 period. The higher match rate in the GWOT-CF is to be expected, since most of those deployments are to locations with no fixed medical treatment facility. Many activations, as defined in the pay file, are likely to be to locations within the United States (e.g., backfilling for active-component soldiers deployed abroad).

[10] The GWOT-CF and PDHA deployment end dates rarely coincide exactly. This is because the PDHA deployment end date is reported by the servicemember, whereas the GWOT-CF deployment end date is reported by the Army. Moreover, servicemembers often complete the PDHA before their deployment has actually ended, and so the end date reported on the PDHA is an estimate. We assume a match if the two end dates are within two months of each other.

below). Employing the same method, we were able to find PDHA forms for 90 percent of eligible reserve-component members.

Correlation Between Post-Deployment Health and UCX Receipt

The PDHA form inquires about a range of physical and mental health conditions. The form asks respondents to report whether their health improved, stayed the same, or worsened during deployment; whether their health currently is excellent, very good, good, fair, or poor; how many days they were on sick call; and whether they spent one or more nights in a hospital. In another section, the form asks respondents to report whether they suffered particular physical symptoms at the time they completed the PDHA. These symptoms include chronic cough, runny nose, fever, weakness, headaches, swollen joints, back pain, muscle aches, numbness, skin disease, redness of eyes, dimming of vision, chest pain, dizziness, difficulty breathing, feeling tired, difficulty remembering, diarrhea, frequent indigestion, vomiting, and ringing of the ears.

The PDHA asks a series of questions about the individual's mental health, including whether the individual feels depressed and whether the individual has or had plans to seek counseling or care for his or her mental health. One question is designed to screen for post-traumatic stress disorder (PTSD). We follow Hoge, Auchterlonie, and Milliken (2006) in defining an individual as having screened positive for PTSD if he or she answers yes to two or more of the following questions: "Have you ever had any experience that was so frightening, horrible, or upsetting that, in the past month, you (1) Have had any nightmares about it or thought about it when you did not want to? (2) Tried hard not to think about it or went out of your way to avoid situations that remind you of it? (3) Were constantly on guard, watchful, or easily startled? (4) Felt numb or detached from others, activities, or your surroundings?

Finally, the form records whether a health care professional has referred the individual for a list of particular conditions including cardiac, combat/operational stress, dental, dermatologic, ear nose and throat, eye, family problems, fatigue, audiology, gastrointestinal, urological, gynecological, mental health, neurological, orthopedic, pregnancy, pulmonary, or other condition.

We use the large number of questions on the PDHA form to create a smaller set of variables that summarize physical and mental health. Table 4.1 lists these variables, their definition, and the percentage of the Army active- and reserve-component personnel in our sample who display these conditions. These percentages vary from as low 8 percent for hospitalization while deployed to as high as 24–28 percent for being referred for any physical condition. More than a quarter of veterans in our sample state that their health worsened while deployed. Depression affects 27 percent of Army active-component veterans and 19 percent of Army reserve-component veterans upon redeployment; 15 and 11 percent of Army active- and reserve-component members in this sample screen positive for PTSD.

The second and third columns of Table 4.2 report the results of bivariate linear probability regressions of UCX receipt on each of the health variables listed in Table 4.1. Overall, the results reported in Table 4.2 indicate a positive correlation between poor post-deployment mental or physical health and the receipt of UCX. This correlation is generally stronger in the reserve components than in the active components. For example, reserve-component members who report that their health worsened while deployed are 5.3 percentage points more likely to claim UCX than reservists who did not, which represents an increase over the mean propensity to claim (17.1 percent) of 31 percent. Active-component members who report that their health

Table 4.1
Post-Deployment Physical and Mental Health Variables Derived from the PDHA

		Mean (percentage)	
Variable	Definition	Active Component	Reserve Component
Poor health	Health is currently poor or fair	13	11
Health worse	Health worsened while deployed	25	25
Four or more days sick	Five days or more on sick call while deployed	22	27
Hospitalized	Hospitalized for one or more nights while deployed	8	8
Any physical symptom	Suffer one or more physical symptoms	16	20
Three or more physical symptoms	Suffer three or more physical symptoms	6	8
Any physical referral	Referred for one or more physical conditions	28	24
Depressed	Over the past two weeks felt depressed or felt little interest or pleasure in doing things	27	19
PTSD	See text	15	11
Sought mental health care	Sought or plan to seek counseling for mental health	9	8
Any mental health referral	Referred for one or more mental health conditions	7	5
Any condition	One or more conditions listed above	65	63

SOURCES: PDHA and RAND UCX database.

worsened while deployed are 3.2 percentage points more likely to claim UCX than active-component veterans who did not, which represents an increase over the mean propensity to claim (24.4 percent) of 13 percent.

Mental health conditions appear to be more strongly correlated with the propensity to claim UCX than are physical conditions. For example, reservists who report being depressed are 6.8 percentage points more likely to claim UCX than reservists who are not. Screening positive for PTSD elevates the likelihood that a reservist will claim UCX by 9.0 percentage points.

To account for the possibility that these bivariate correlations reflect deployment duration (the chance of poor post-deployment health increases with deployment duration)[11] and pay grade (lower-ranking servicemembers may be more likely to claim UCX and report poor health), we implement a series of multivariate linear probability regressions of UCX receipt on each of these health conditions, pay grade (E0–E3, E4, E5, E6–E9, W1–W5, O1–O3, and O4–O6), and deployment duration (as defined in Figures 4.1 and 4.2). The fourth and fifth columns of Table 4.2 report coefficient estimates on the health variables derived from these multivariate regressions. As can be seen, the correlation between poor post-deployment mental and physical health and UCX receipt remains after controlling for deployment duration and pay grade.

[11] For example, the probability of reporting any condition increases by 4, 13, and 15 percentage points for reservists activated 7–9, 10–15, and 16 or more months, respectively, relative to reservists activated 4–6 months. The correlation between deployment duration and poor post-deployment health for active-component members is much weaker.

Table 4.2
The Effect of Post-Deployment Health on UCX Receipt: Bivariate and Multivariate Estimates, by Army Component

	Bivariate Effect		Multivariate Effect	
Explanatory Variable	Active Component	Reserve Component	Active Component	Reserve Component
Poor health	0.032	0.053	0.028	0.049
	(0.011)	(0.006)	(0.011)	(0.006)
Health worse	0.026	0.046	0.027	0.052
	(0.008)	(0.004)	(0.008)	(0.004)
Four or more days sick	0.037	0.045	0.037	0.04
	(0.009)	(0.004)	(0.009)	(0.004)
Hospitalized	0.009	0.027	0.003	0.025
	(0.013)	(0.006)	(0.013)	(0.006)
Any physical symptom	0.005	0.023	0.007	0.024
	(0.010)	(0.004)	(0.009)	(0.004)
Three or more physical symptoms	0.026	0.041	0.027	0.041
	(0.015)	(0.007)	(0.015)	(0.006)
Any physical referral	0.019	0.017	0.02	0.023
	(0.008)	(0.004)	(0.008)	(0.004)
Depressed	0.031	0.068	0.024	0.052
	(0.008)	(0.005)	(0.008)	(0.005)
PTSD	0.024	0.09	0.02	0.076
	(0.010)	(0.006)	(0.010)	(0.006)
Sought mental health care	0.041	0.074	0.037	0.063
	(0.013)	(0.007)	(0.013)	(0.007)
Any mental health referral	0.012	0.064	0.011	0.052
	(0.014)	(0.008)	(0.014)	(0.008)
Any condition	0.026	0.049	0.025	0.046
	(0.007)	(0.003)	(0.007)	(0.003)
Mean UCX receipt	0.244	0.171	0.244	0.171
Obs.	15,090	51,924	15,090	51,924

SOURCES: PDHA and RAND UCX database.

NOTES: Each cell reports the coefficient estimate from separate linear probability models of the receipt of UCX in the second quarter of 2005 on the health variables listed in the first column. The second and third columns report estimates from bivariate regressions. The third and fourth columns report estimates from multivariate regressions where the additional variables are pay grade and deployment duration (defined categorically in the text above). Standard errors are in parentheses. See the text for sample restrictions.

One reason why the correlation between poor post-deployment health and UCX receipt could be higher for reservists than for active-component members is that reservists largely qualify for UCX benefits based on deployment, whereas active-component members do not. Thus, for reservists, the negative health effects of deployment are reported quite close to the period of eligibility for UCX. Active-component members, on the other hand, could have been deployed many months or even years prior to separating from military service and becoming eligible to receive UCX.

Interpreting the Correlation Between Post-Deployment Health and UCX Receipt

We caution here that we cannot conclude from Table 4.2 that poor health causes veterans to claim UCX at higher rates than they otherwise would. It is possible that veterans who are predisposed to claiming UCX are also predisposed to reporting that they are in poor physical or mental health. So, whether it is poor health or some unobservable characteristic of veterans that causes them to claim UCX is unknown. However, even if we interpret these correlations between poor health and UCX receipt as purely causal, it is unlikely that rising rates of poor health can account for much of the increase in UCX claim rates.

Consider the Army active components. Claim rates for those veterans separating in the first two quarters of 2002 and 2005 increased by 3 percentage points. Active-component veterans who suffered any injury listed in Table 4.1 were 2.5 percentage points more likely to receive UCX than those who were not injured. About 65 percent of eligible active-component members suffered one or more of these conditions. But 56 percent of eligible active-component members had no previous deployment. Thus, at best, rising injury rates could explain 0.7 percentage points of the increase in the active-component UCX claim rate. This estimate, though, is likely to be too high, since surely some fraction of veterans eligible to receive UCX in the second quarter of 2002 also were in poor health following deployment (the PDHA data are incomplete prior to spring 2003, so we cannot estimate this fraction directly), and so the increase in the prevalence of poor health in the eligible population between 2002 and 2005 is likely considerably less than 65 percent.

Now consider the Army reserve components. Claim rates for reservists completing 90 or more days of active-duty service in the first two quarters of 2002 and 2005 increased by 11 percentage points. Reserve-component veterans who suffered any adverse health condition listed in Table 4.1 were 4.6 percentage points more likely to receive UCX than those who did not. About 63 percent of eligible active-component members suffered one or more of these conditions. Thus, at best, rising injury rates could explain 2.9 percentage points of the observed 11-percentage-point increase in the reserve-component UCX claim rate. As with active-component veterans, though, this estimate is likely to be too high, since some fraction of reservists eligible to receive UCX in the second quarter of 2002 also were in poor health following deployment, so the increase in the prevalence of poor health in the eligible reserve population between 2002 and 2005 is likely considerably less than 63 percent.

CHAPTER FIVE

The Pre- and Post-Activation Employment Experiences of UCX Recipients

The analyses of the previous chapters suggest that more than half of the increase in the Army reserve-component UCX caseload is attributable to an increase in UCX claim rates. We have argued that changes in the civilian labor market cannot explain rising UCX claim rates. Why, then, is such a large fraction of Army reservists now claiming UCX (about 13 percent of those eligible to claim in the seven months prior to the end of the second quarter of 2005)? The similarly high claim rates (19 percent) of Army active-component members is perhaps more easily understood. Most of these veterans have no significant civilian labor market experience, and it is reasonable to assume that many will require some time to find civilian employment that takes advantage of the skills they developed in the military. A large fraction of reservists, on the other hand, are employed in the civilian labor market prior to being activated and, under the Uniformed Services Employment and Reemployment Rights Act (USERRA) of 1994 (P. L. 103-353), have the right to return to that job following activation. Since UCX generally requires applicants to accept suitable employment, the availability of a USERRA-protected job should make UCX eligibility far less likely.[1]

In this chapter, we employ data from the Status of Forces Survey of Reserve Component Members (SOFS-R) to describe the pre- and post-activation employment experiences of UCX recipients. We continue to focus on the Army reserve components in this chapter, since they account for the bulk of the increase in the UCX caseload and because, as we argue in the concluding chapter, the role UCX should play in assisting reservists return to the civilian labor market is perhaps less obvious.

The SOFS-R

The Status of Forces Surveys, administered by DMDC, are a suite of periodic surveys of active- and reserve-component members and DoD civilian employees. These surveys are designed to track opinions, attitudes, and experiences of DoD military and civilian personnel. We employ the May 2007 SOFS-R for the purposes of this chapter. The SOFS-R was conducted online

[1] A determination of whether refusal of suitable employment renders an individual ineligible for UCX is made under state law on a case-by-case basis. Even if a reservist has access to a USERRA-protected job, other considerations, some of which vary state to state, might nonetheless result in a determination of UCX eligibility.

and was designed to be representative of individuals actively serving in the Selected Reserves.[2] All analyses reported in this chapter employ survey weights designed to correct for stratified sampling and disproportionate survey nonresponse.[3]

The SOFS-R surveys respondents on a wide range of topics. We focus our analyses on the battery of questions in the June 2007 SOFS-R concerning pre- and post-activation employment. We limit our analysis sample to 5,119 respondents who were members of the Army National Guard and Reserve, who had been deployed for 90 or more days since September 11, 2001, but who were not currently activated.[4]

Pre-Activation Employment and UCX

About 65 percent of our analysis sample were employed in the month prior to being activated; 94 percent of these individuals worked for an employer, and 6 percent were self-employed or worked for a family business. We estimate that about 11 percent of those who were not employed in the month prior to activation were part-time students and another 12 percent were full-time students.[5] The balance, about 12 percent, were neither working nor in school.

Perhaps not surprisingly, employment following activation is less likely among individuals who were not employed in the month preceding their most recent activation. Among those who were not employed prior to activation, 77 percent reported not working for pay in the three months following deactivation. Among those who were employed prior to activation, only 16 percent did not work for pay following deactivation. UCX receipt is somewhat higher among those not employed prior to activation, but the difference in UCX receipt across the two groups is perhaps not as much as we might expect, given the difference in employment rates; 11 percent and 9 percent of those not employed and employed prior to activation, respectively, reported receiving UCX in the three months following activation.[6]

---

[2] Reservists who had less than six months of service when the survey was conducted or who were of flag rank when the sample was drawn (six months prior to the survey) were excluded from the survey sampling frame. Reservists who were selected to participate in the survey were notified by mail one month before the survey was actually administered, and second notifications were issued via email within 24 hours after the questionnaire was posted on the Web site. Sampled individuals who did not return a completed survey were sent up to six reminder emails and three reminder letters. For more information about the SOFS-R, please refer to DMDC (2005). Documentation for the June 2007 SOFS-R was not available at the time of publication.

[3] The May 2007 SOFS-R had a response rate of about 32 percent. Martorell, Klerman, and Loughran (2008) report that survey weights were effective in correcting for nonresponse bias in civilian and military earnings in the 2004 and 2005 May SOFS-R. Whether the survey weights are effective in correcting for nonresponse bias in other survey items (such as the unemployment questions we examine here) is unknown.

[4] This subsample represents about one-quarter of survey respondents who provided a valid component.

[5] These estimates of the percentage in school are based on answers to questions about the current employment of individuals not activated in the past 24 months. We make no attempt to adjust for the likelihood of activation.

[6] The SOFS-R allows respondents to specify whether they received UI or whether they received UCX. Among those receiving UI or UCX, 81 percent stated they received UI, and 19 percent stated they received UCX. It seems likely that many servicemembers mistakenly classified their benefit as coming from UI, since UI is a more familiar term for unemployment benefits. Also, the question offered UI as the first item and UCX as the second item, and respondents may have simply selected the first item without much thought as to which program in fact applied.

Post-Activation Employment and UCX

The SOFS-R indicates that 59 percent of reservists who received UCX in the three months following deactivation were employed in the month prior to being activated. Since this employment relationship is protected by USERRA, these reservists should generally have access to suitable employment when they return, which should make them far less likely to qualify for UCX. Only the self-employed, individuals whose jobs were eliminated (either because the employer ceased operations or the position itself was eliminated), and those who quit or were laid off from their jobs before activation are not protected by USERRA.[7] Some employers willfully disregard USERRA and refuse reemployment, but, as we will see below, this is relatively rare. So, the fact that the majority of UCX recipients were employed prior to activation is puzzling.

The SOFS-R asks reservists who were employed prior to activation whether they experienced particular problems returning to their pre-activation job. These problems are listed in Table 5.1 along with the fraction of reservists answering that they had experienced each problem, by whether they received UCX. Overall, 29 percent of UCX recipients stated that they experienced at least one of the problems listed in Table 5.1. Thirty-one percent of reservists who reported having experienced at least one of these problems also stated that they had contacted the National Committee for Employer Support of the Guard and Reserve (ESGR) to seek assistance in resolving these problems.[8]

Table 5.1
Ex-Servicemembers Reported Reemployment Problems, by UCX Receipt

	Received UCX (%)	
Reemployment Problem	**Yes**	**No**
Denied promotion that would have been granted if not for military service	8	12
Loss of seniority, seniority-related pay, or seniority-related benefits	13	16
Denied the same job as held prior to activation	16	11
Employer could not accommodate reemployment claims	12	3
Demotion to lower position than held prior to activation	7	7
Military service considered a break in employment for pension benefit purposes	6	16
Failed to receive immediate reinstatement of employer provided health insurance	2	7
Reasonable efforts not made to refresh or upgrade skills to enable you to qualify for reemployment	6	7
Termination without cause (31–80 days after demobilization)	4	1
Termination without cause (more than 180 days after demobilization)	4	2

SOURCE: 2007 SOFS-R.

[7] USERRA protection is not extended in a few other circumstances as well. See Employer Support of the Guard and Reserve's (ESGR's) Web page on USERRA (Employer Support of the Guard and Reserve, n.d.).

[8] The ESGR is a DoD organization whose purpose is to promote cooperation and understanding between reserve-component members and their civilian employers and to assist in the resolution of conflicts arising from an employee's military commitment.

The SOFS-R asks reservists who were employed prior to activation, but not employed in the three months following deactivation, why they did not return to their pre-activation employer. We tabulate the response to that question in Table 5.2 by whether reservists received UCX benefits in the three months following deactivation. Among UCX recipients, 5 percent stated that their pre-activation employer went out of business, 33 percent stated that pre-activation employer circumstances changed, and 14 percent stated that their pre-activation employer did not give them prompt reemployment (respondents could list more than one reason, and so the column percentages sum to more than 100 percent). In these situations, USERRA may not provide protection against involuntary separation from a pre-activation job.

On the other hand, 16 percent of UCX recipients stated that they disliked their pre-activation job, 24 percent stated that they decided to attend school, and 51 percent stated that they needed a break. In principle, the UCX program should not provide benefits to individuals who fail to return to a pre-activation job for these reasons alone. Overall, 47 percent of UCX recipients cited involuntary reasons for not returning to a pre-activation job, and 61 percent cited voluntary reasons; about 26 percent cited only involuntary reasons, and 40 percent cited only voluntary reasons.

Table 5.2
Ex-Servicemembers's Reasons for Not Returning to Pre-Activation Employer, by UCX Receipt

	Received UCX (%)	
Reason Did Not Return to Pre-Activation Employer	**Yes**	**No**
Pre-activation employer went out of business	5	5
Changes in pre-activation employer circumstances (e.g., layoff occurred, facilities closed, ownership of company changed, contract ended)	33	13
Pre-activation employer did not give me prompt reemployment (i.e., within two weeks of return from military service)	14	7
I disliked my pre-activation job	16	10
I was recuperating from an illness/injury	10	13
I decided to attend school	24	17
I needed a break after activation	51	57
Other	21	39
Involuntary reason	47	29
Voluntary reason	61	64
Only involuntary reason	26	13
Only voluntary reason	40	48

SOURCE: 2007 SOFS-R.

CHAPTER SIX

Conclusion

This final chapter serves three roles. First, we review our findings and their implications for the civilian labor market for veterans. Second, we consider the implication of our findings for the UCX program. Finally, we discuss several directions for future research on the civilian labor market experiences of recent veterans.

Have Civilian Labor Market Conditions for Veterans Deteriorated?

This study was motivated by concern that sharp increases in the UCX caseload reflected a worsening civilian labor market for veterans. The findings reported in this document suggest that this concern is misplaced. The increase in the UCX caseload is due to the large increase in UCX usage among reservists. That increase in UCX usage is due to more reservists serving on active duty and, therefore, becoming eligible for UCX, and to an increase in the fraction of eligible reservists claiming UCX.

While the increase in the number of reservists eligible to receive UCX is unsurprising, the increase in the claim rate is potentially a cause for concern. Two pieces of evidence, though, suggest that the increase in the reserve-component UCX claim rate most likely does not reflect a deterioration of the civilian labor market for veterans. First, between 2002 and 2005, the civilian labor market generally improved and, although the unemployment rate of veterans ages 20–24 increased between 2003 and 2005, Savych, Klerman, and Loughran (2008) argue that this trend most likely reflects sampling variation rather than real changes in the veteran youth unemployment rate.

Second, the Army and Air Force active-component UCX claim rates increased only slightly between 2002 and 2005. If the civilian labor market was deteriorating for veterans during this time period, we would expect active-component UCX claim rates to increase perhaps even more than reserve-component UCX claim rates, since reservists, by their very nature, are more integrated into the civilian labor market.

If the civilian labor market for veterans has not deteriorated, why has the reserve-component claim rate risen? Our analyses suggest that part of the explanation lies in longer deployments. Army reserve-component UCX claim rates are positively correlated with deployment duration, and deployment duration increased markedly over our study period. While the probability of injury increases with deployment duration, and poor post-deployment health is positively correlated with UCX claim rates, we argue that these correlations cannot account for much of the increase in UCX claim rates. Thus, something else about longer deployments must be to blame for rising UCX claim rates.

One possibility is that longer deployments make it more difficult for predeployment civilian employers to accommodate reemployment. Among reservists surveyed in the SOFS-R who worked prior to deployment, 37 percent stated that they had some type of reemployment problem, and 16 percent of these individuals stated that they had reported these problems to ESGR. Among those reservists who did not work in the three months following activation, 25 percent stated that their pre-activation employer had gone out of business, that their particular job had been eliminated, or that their employer had denied them prompt reemployment.

Another possibility is that reservists desire some time off following deployment prior to returning to their civilian employer. In the SOFS-R, among reservists who did not work in the three months following deployment, but were employed in the month prior to deployment, 55 percent stated that they did not work because they "wanted a break."

Finally, longer deployments might cause some reservists to reconsider their civilian careers. During their deployment, they may have learned new skills and been exposed to new people with different perspectives and life experiences that could cause them to consider alternative labor market opportunities when they return home. In the SOFS-R, 12 percent of reservists reported that they did not work following deployment because they disliked their predeployment job, and 19 percent reported that they decided to attend school.

Implications for Policy

USERRA provides for up to three months of protected leave following deployment, and many reservists can draw on paid leave that they have accumulated while serving on active duty to support time off from civilian employment, whether it be to recuperate from a stressful deployment or to consider alternative employment.[1] An open question, though, is whether UCX should provide additional income support to reservists who choose not to return to civilian employment for these reasons.

The UI program is intended to support individuals with a demonstrated attachment to the labor force, but who have been involuntarily separated from a job and are actively searching for new work. For active-component members, implementation of the UCX program seems reasonably consistent with these UI objectives. Veterans separating from the active components typically will have served for a minimum of three years, and very often their military service will have been their first significant employment. Thus, these veterans typically will not have established a civilian career or have a specific civilian job waiting for them when they separate. UCX supports their transition to the civilian labor market.

The extension of the UI program to reservists is less obvious. Reservists are typically older, and many have established civilian careers. Moreover, their period of active-duty service is comparatively short (usually 18 months or less), so it is reasonable to expect that most will have a civilian job, typically protected by USERRA, to which they can return. However, tabulations of SOFS-R data suggest that 57 percent of UCX recipients were employed in the month prior to deployment and that 40 percent of these individuals chose not to return to that job for exclusively voluntary reasons.

---

[1] Reservists who serve 180 or more days on active duty must apply for reemployment within 90 days of completing service. For reservists serving 30–179 days, application for reemployment must be submitted within 14 days of completing service.

These facts suggest that policymakers might reconsider the objectives of the UCX program.[2] In its current form, the UCX program is intended to mirror the civilian UI program. As such, reservists who have civilian jobs to which they can return should not be eligible for UCX. Enforcing this requirement, however, may entail developing new procedures that could be followed by state workforce agency personnel. For example, UCX regulations might require state workforce agencies to verify whether reservists were employed prior to activation and, if so, to determine whether the right to return to that job is protected by USERRA. Reservists claiming UCX might be required to sign a statement that they do not have a USERRA-protected job to which they could return. Reservists who do not return to a USERRA-protected job would be eligible to receive UCX only under special circumstances, such as when the job was eliminated or their employer refused them reemployment.

Current DoL guidance, as specified in Unemployment Insurance Program Letter (UIPL) No. 27-06 issued in August 2006, suggests that states should be verifying availability to work for a pre-activation employer. The DoL guidance includes the following two questions and answers:

> **Question:** If UCX claimants fail to contact their last employer to exercise their employment rights under the Uniformed Services Employment and Reemployment Rights Act of 1994 (USERRA), how must states adjudicate any resulting issue(s)?
>
> **Answer:** The state should conduct appropriate fact-finding to determine if the claimant is able and available for work and to determine if the claimant has refused suitable work as defined by state law. (UCX claimants are subject to the same able and available requirements of state law as regular unemployment claimants, including any work search requirements.)
>
> **Question:** When individuals file an initial claim or reopen a claim after a military discharge and they plan to return to full-time employment in the near future with a pre-service employer, does the state have to take any special action(s)?
>
> **Answer:** The action required depends on the unemployment insurance (UI) law of the individual state. Some states waive the work search requirement if claimants are job-attached and other states require a work search regardless of job attachment. Individuals filing UCX claims are subject to the same work search requirements as those under the state UI law and the state must adjudicate any issue(s) as it would on any other claim.

This guidance, however, is not consistent with language in the UCX Handbook (DoL ETA, 1994, Chapter 4, 7.a). At the time this report was written, DoL was in the process of revising the handbook to be consistent with UIPL No. 27-06.

We should note, though, that a requirement of returning to a pre-activation job creates certain inequities and perhaps perverse incentives. Reservists who were not employed prior to activation are generally entitled to receive UCX just like active-component members, providing that they are able and available to work. However, a reservist whose pre-activation job was protected by USERRA would generally need to demonstrate that that job was not available to him following activation or was otherwise unsuitable in order to qualify for UCX. A forward-

[2] We use "policymakers" generically here to refer to federal and state agencies with authority to enforce and interpret UCX and USERRA statutes and regulations. Moreover, some changes might require revisions to the UCX and USERRA statutes.

looking reservist who did not like his pre-activation job, therefore, might quit that job prior to activation in order to improve his chances of receiving UCX following activation.

On the other hand, one could argue that UCX should not be so narrowly targeted. DoD, DoL, and DVA operate a number of programs designed to help reservists transition to civilian employment following activation. However, none of these programs provide income support for reservists who want to make a career change. Whether providing such income support is desirable in the context of overall reserve compensation and whether UCX is the right vehicle with which to deliver such income support are open questions.

Directions for Future Research

This study suggests a number of directions for future research that would deepen our understanding of the reasons why veterans choose to claim UCX benefits. First, the data on UCX utilization for this study were limited. For the Army and Air Force, data covered two quarters. For the Navy and Marine Corps, the data covered only one quarter. The limited coverage of our data was due to the expense of transcribing paper records into an electronic database and limitations in service recordkeeping. Obtaining electronic data on UCX receipt from individual states or from the Federal Claims Control Center would allow analyses over more periods and more services, as well as an improved characterization of benefit duration.

Second, we would like to better understand the interaction between UCX receipt, interstate variation in UI/UCX policy, service characteristics (e.g., rank, Military Occupation Specialty, deployment, combat experiences, and reasons for leaving active duty), pre-activation employment characteristics (of reservists), labor market conditions, and subsequent labor market outcomes. Subsequent labor market outcomes of interest might include speed of return to work, return to pre-activation employer, and changes in wages and earnings. The data available for this study would not support analyses of that form, but other data, such as the Census Longitudinal Employer-Household Dynamics (LEHD) database, could.

Finally, the correlation between post-deployment health and UCX receipt described here is deserving of further investigation. Of particular importance is employing data and methods capable of isolating the causal effect of injury on civilian labor market outcomes.

APPENDIX A

A Brief Description of the UI and UCX Programs

This appendix provides additional detail on the UI and UCX programs.

Unemployment Insurance

Unemployment Compensation (UC) is a social insurance program created by the Social Security Act of 1935. UC, or Unemployment Insurance (UI), as it is usually called, is a joint federal-state program. The federal government sets broad program rules. With wide latitude, states choose specific program details.

This section lays out the broad federal rules and the specific state implementation. Unless otherwise noted, the discussion here is based on the official DoL interstate analysis, as of January 2006 (DoL ETA, 2006a).

Labor Force "Attachment"

The UI program provides benefits to currently unemployed individuals who were recently "attached" to the labor market. The details of the definition of "attached" (i.e., whether one qualifies for any check) and of the benefit (i.e., how large is the check) vary between states.

Labor market attachment and benefits for UI are both assessed from wages earned within a time frame of a claimant's employment history known as the *base period*. The base period in nearly every state is the first four of the last five calendar quarters preceding the calendar quarter in which the claim is filed. Because this definition can potentially overlook up to six months of a claimant's most recent employment history, some states allow the use of an *alternative base period* (usually the last four completed calendar quarters). Some states allow the use of an *extended base period*, which looks at wages and employment from earlier quarters, to adjust for claimants with prolonged illness or injury.

A UI claim that satisfies one of the criteria will establish a *benefit year*, which is defined as a one-year or 52-week period during which a claimant may receive benefits. In nearly every state, the benefit year begins the week a valid claim is made.[1]

States use a variety of formulas to test for monetary eligibility from the base period. About half of the states use some version of the *Multiple of High-Quarter Wages Method*. This method requires the applicant's wages to total some designated amount in the quarter of highest earnings, and the total earnings in the base period to be some multiple of the high-quarter

[1] The two exceptions are Arkansas, where the benefit year begins with the first day of the quarter in which the claim is filed, and New York, where the benefit year is a 53-week term beginning with the effective date of a valid claim.

wages. For example, a claimant in Louisiana must have high-quarter wages of at least $800 and must have earned a total of at least 1.5 times the high-quarter wages ($1,200) in the base period. States using this method typically impose a condition of employment in at least two quarters of the base period to demonstrate labor attachment.

Many of the remaining states use the *Multiple of Weekly Benefit Amount Method*. Determination of monetary entitlement under this method begins with computation of the *weekly benefit amount* (described below). The claimant then needs to have earned some designated multiple of the weekly benefit amount in the base period to qualify for UI. For example, Mississippi's formula is 40 times the weekly benefit amount in the base period and 26 times the weekly benefit amount in the high quarter. Again, states using this method will usually require work in more than one quarter of the base period.

A flat qualifying amount of wages or weeks/hours of employment is the sole criterion in a few states, while in other states it is used as an alternative to or in combination with another method. The qualifying amount is sometimes based on the state minimum wage and the statewide average weekly or annual wage. For example, Iowa requires 1.25 times high-quarter wages for the base period plus high-quarter wages of at least 3.5 percent of the average annual wage, and in Ohio, the monetary requirement is 20 weeks of work with wages averaging 27.5 percent of the statewide average weekly wage.

The UI Benefit

States use one of four different schemes to calculate the *weekly benefit amount*. Over half calculate the weekly benefit from earnings in the high quarter. Suppose, for example, that a claimant earns $3,000 in the high quarter of the base period, and the state's UI policy is to replace 50 percent of the weekly wage. Each quarter is 13 weeks, so the weekly benefit would be $3,000 times 1/26, or approximately $115 per week. Frequently, states decrease the denominator to account for unemployment in the high quarter or to increase the weekly benefit if a claimant has dependents.

The multiquarter method is similar but computes the weekly benefit by taking some fraction of the total or average wages earned over more than one quarter.

A third method is to equate the base period earnings to an annual wage, some percentage of which becomes the weekly benefit. Typically, the proportions are weighted to give low-wage earners a higher percentage of their annual wage as a weekly benefit.

Duration of Benefits

How long the payments will continue also varies among states. A minority of states provide benefits for a uniform duration of 26 weeks within any benefit year, which is the maximum duration for most of the United States. However, one should not misinterpret uniform duration states as being more generous than variable duration states, because the former's eligibility requirements are often more stringent. Variable duration states usually derive the maximum entitlement for any benefit year from the lesser of 26 times the weekly benefit amount and some fraction of the base period wages. Other states with a variable duration of benefits set the maximum entitlement at some fraction of the total weeks worked during the base period. In either case, the maximum entitlement is then divided by the weekly benefit amount to determine the duration of benefits.

Other Details

Claimants who are partially unemployed may still be eligible for unemployment compensation. Most states define *partial unemployment* as a week of less-than-full-time work in which the wages earned were less than the weekly benefit amount for the state. The weekly benefit amount for the partially unemployed is adjusted by subtracting some amount, but never all, of the wages earned, so as not to discourage partial employment. Some states, such as Hawaii, disregard a fixed amount of earnings ($50), while others, such as Arizona, disregard earnings up to some percentage of the claimant's weekly benefit amount.

In all but 14 states, there is a one-week waiting period between application and first check at the beginning of the first benefit year. Only New York increases the waiting period to two weeks in cases of partial unemployment.

A claimant's weekly benefit amount may be supplemented by a *dependents allowance* if he or she is deemed a crucial means of support to certain family members. Most states have no dependents allowances, and for those 13 that do, there are marked differences in their definitions of dependents and in their dependents allowances.

Nonmonetary Eligibility Criteria

Crucial for our analysis of UCX, nonmonetary eligibility criteria make up another important category of unemployment compensation laws, but they tend to be more straightforward and less variable than entitlement policies. In part, the greater uniformity is due to federal law, which mandates that, in addition to demonstrating labor market attachment, UI applicants be able to show that they are: (1) unemployed through no fault of their own and (2) able and available to work. Most states also require that applicants be actively seeking work. In addition, a plethora of nonmonetary eligibility laws pertain to UI disqualification of persons "voluntarily" leaving work (e.g., due to job dissatisfaction or a labor dispute).

Unemployment Compensation for Ex-Servicemembers (UCX)

Having described the civilian UI program, we now turn to UCX. Established as a permanent federal program in 1958, UCX makes ex-military personnel eligible for UI. As its name implies, the program was designed to provide income assistance to former active-duty members of the military as they search for work.

At its founding, servicemembers were eligible for the same benefits as civilians, but without the requirement of involuntary separation. The Miscellaneous Revenue Act of 1982 (P. L. 97-362) delayed payment from the second to the fourth week after discharge and cut the maximum duration of benefits from 26 weeks to 13 weeks. However, as part of the drawdown, the Emergency Employment Compensation Act of 1991 (P. L. 102-107) stipulated that the UCX program would, for the most part, abide by the applicable state laws governing UI for other civilians. This legislation had the effect of returning the first payable period to the second week after discharge and the maximum duration of benefits to 26 weeks.

Because military service is federal employment, and an exceptional form of employment in other respects, the UCX program has a larger federal role and its own set of eligibility criteria. Specifically, in accordance with a cooperative agreement between DoL and DoD, military personnel are apprised of the UCX program upon issuance of their DD Form 214/215, a Certificate of Release or Discharge from Active Duty. The DD Form 214/215 details the period of service and the reason for discharge; ex-servicemembers are asked to refer to this document when applying for UCX benefits. States then verify the DD 214/215 information with the FCCC, which maintains copies of these forms.

As mentioned previously, UCX has very specific eligibility requirements. To qualify for UCX, an individual must have completed the first full term of active military service he or she agreed to serve.[2] Furthermore, he or she must not have been discharged dishonorably or, if an officer, have left for the good of the service.[3] Reservists face the added requirement of a minimum of 90 consecutive days on active duty.

It is crucial to note that, under USERRA, ex-servicemembers have a legal right to reemployment with their pre-service employers, regardless of employment sector and regardless of whether they were inducted, activated from the reserves, or enlisted into military service. Failure to resume employment with the pre-service employer, when it is feasible to do so, will typically disqualify a claimant from receiving UCX benefits, although such a determination is made on a case-by-case basis following state law. By statute, states are explicitly prohibited from applying their own UI eligibility laws with respect to reasons for separation from active military service or failure to seek or accept employment with a pre-service employer.

Military branches reimburse the federal government for the UCX benefits distributed to ex-servicemembers through the state agencies. Claims submitted by the states are paid out of the Treasury Department's Federal Employees Compensation Account, and the records of these claims are sent to the military branches for billing purposes on a quarterly basis.

Under ordinary UI, workers are generally covered by the UI law of the state in which the work was performed. When ambiguities arise, such as they might in the case of a traveling salesman, for example, "localization of work" provisions are used to assign work to a particular state.[4] Because the military is part of the federal government and its members may perform their duties outside the United States, there is an inherent difficulty in assigning to any particular state the wages earned while in its employ. To circumvent this problem, federal law grants ex-servicemembers the right to file their initial claim in any state where they are physically located. However, the statute is not uniformly interpreted: Some states require that the

---

[2] An example in ETA Handbook No. 384 (DoL ETA, 1994) on p. II-5 states that the UCX service requirement is met by a person who is honorably discharged after 60 days for a service-incurred injury, but not by an Academy cadet who agreed to five years of active duty but is discharged for an unsatisfactory reason even after more than 90 consecutive days of active duty.

[3] The full text, from ETA Handbook No. 384 (DoL ETA, 1994), p. II-2, reads that a claimant must have been

1. on active duty in the Armed Forces or the Commissioned Corps of the National Oceanic and Atmospheric Administration. However, members of the National Guard or other Reserve Component of the Armed Forces must have been on continuous active duty in a reserve status for 90 days or more
2. discharged or released under honorable conditions and, if an officer, did not resign for the good of the service
3. discharged or released after completing the first full term of active military service which the individual initially agreed to serve, unless the individual was discharged or released before completing such term of active service for one of the following reasons: a) The convenience of the Government under an early release program; b) Because of medical disqualification, pregnancy, parenthood, or service-incurred injury or disability; c) Because of hardship; or d) Because of personality disorders or inaptitude after having continuously served for 365 days or more.

[4] If a worker's services are not localized to any state, his work may be assigned to the state of his base of operations, the state of his employer's base of operations, or the state in which he resides.

claimant have a verifiable mailing address within the state, while other states permit patently transient applicants.[5] Enforcement of the federal law is complicated by the policies of some states that allow filing claims by telephone or through the Internet.

Federal law also enables ex-servicemembers with base period earnings from a source other than the military to have those wages considered in eligibility and benefit determinations. Ex-servicemembers with additional base period earnings can file a *joint claim*, and the joint claim need not be filed in the state in which those additional wages were earned. In filing an *interstate claim*, however, the claimant is still subject to the UI laws of the state in which the claim was filed.

---

[5] Keith Ribnick, Unemployment Insurance Program Specialist with the Department of Labor, personal communication, January 8–10, 2007. According to Ribnick, the strictness of this policy is a matter of the state's interpretation of the term "physically located." Technology has allowed remote filing of claims and introduced new challenges to the system in ensuring that claims are appropriately filed. A number of states have addressed the problem of location through telephone identification systems and Internet protocol (IP) addresses.

APPENDIX B

Form DD 2796

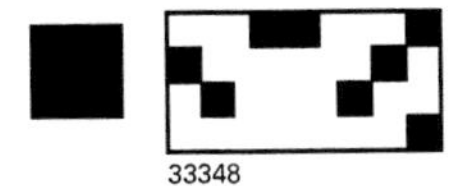

POST-DEPLOYMENT **Health Assessment**

Authority: 10 U.S.C. 136 Chapter 55. 1074f, 3013, 5013, 8013 and E.O. 9397

Principal Purpose: To assess your state of health after deployment outside the United States in support of military operations and to assist military healthcare providers in identifying and providing present and future medical care to you.

Routine Use: To other Federal and State agencies and civilian healthcare providers, as necessary, in order to provide necessary medical care and treatment.

Disclosure: (Military personnal and DoD civilian Employees Only) Voluntary. If not provided, healthcare WILL BE furnished, but comprehensive care may not be possible.

INSTRUCTIONS: Please read each question completely and carefully before marking your selections. Provide a response for each question. If you do not understand a question, ask the administrator.

Demographics

Last Name

First Name **MI**

Name of Your Unit or Ship during this Deployment

Today's Date (dd/mm/yyyy)

Social Security Number

DOB (dd/mm/yyyy)

Date of arrival in theater (dd/mm/yyyy)

Date of departure from theater (dd/mm/yyyy)

Gender
- O Male
- O Female

Service Branch
- O Air Force
- O Army
- O Coast Guard
- O Marine Corps
- O Navy
- O Other

Component
- O Active Duty
- O National Guard
- O Reserves
- O Civilian Government Employee

Pay Grade

O E1	O 001	O W1
O E2	O 002	O W2
O E3	O 003	O W3
O E4	O 004	O W4
O E5	O 005	O W5
O E6	O 006	
O E7	O 007	O Other
O E8	O 008	
O E9	O 009	
	O 010	

Location of Operation
- O Europe
- O SW Asia
- O SE Asia
- O Asia (Other)
- O Australia
- O Africa
- O Central America
- O Unknown
- O South America
- O North America
- O Other

To what areas were you mainly deployed:
(mark all that apply - list where/date arrived)
- O Kuwait
- O Qatar
- O Afghanistan
- O Bosnia
- O On a ship
- O Iraq
- O Turkey
- O Uzbekistan
- O Kosovo
- O CONUS
- O Other

Name of Operation:

Occupational specialty during this deployment (MOS, NEC or AFSC)

Combat specialty:

Administrator Use Only

Indicate the status of each of the following:

Yes	No	N/A	
O	O	O	Medical threat debriefing completed
O	O	O	Medical information sheet distributed
O	O	O	Post Deployment serum specimen collected

33348

DD FORM 2796, APR 2003 PREVIOUS EDITION IS OBSOLETE. ASD(HA) APPROVED

RAND *TR588-AppB.1*

Please answer all questions in relation to THIS deployment

1. Did your health change during this deployment?

O Health stayed about the same or got better
O Health got worse

2. How many times were you seen in sick call during this deployment? [|] No. of times

3. Did you have to spend one or more nights in a hospital as a patient during this deployment?

O No
O Yes, reason/dates: ________________________________

4. Did you receive any vaccinations just before or during this deployment?

O Smallpox (leaves a scar on the arm)
O Anthrax
O Botulism
O Typhoid
O Meningococcal
O Other, list: ____________________
O Don't know
O None

5. Did you take any of the following medications during this deployment?
(mark all that apply)

O PB (pyridostigmine bromide) nerve agent pill
O Mark-1 antidote kit
O Anti-malaria pills
O Pills to stay awake, such as dexedrine
O Other, please list ____________________
O Don't know

6. Do you have any of these symptoms now or did you develop them anytime during this deployment?

No	Yes During	Yes Now	
O	O	O	Chronic cough
O	O	O	Runny nose
O	O	O	Fever
O	O	O	Weakness
O	O	O	Headaches
O	O	O	Swollen, stiff or painful joints
O	O	O	Back pain
O	O	O	Muscle aches
O	O	O	Numbness or tingling in hands or feet
O	O	O	Skin diseases or rashes
O	O	O	Redness of eyes with tearing
O	O	O	Dimming of vision, like the lights were going out
O	O	O	Chest pain or pressure
O	O	O	Dizziness, fainting, light headedness
O	O	O	Difficulty breathing
O	O	O	Still feeling tired after sleeping
O	O	O	Difficulty remembering
O	O	O	Diarrhea
O	O	O	Frequent indigestion
O	O	O	Vomiting
O	O	O	Ringing of the ears

7. Did you see anyone wounded, killed or dead during this deployment?
(mark all that apply)

O No O Yes - coalition O Yes - enemy O Yes - civilian

8. Were you engaged in direct combat where you discharged your weapon?

O No O Yes (O land O sea O air)

9. During this deployment, did you ever feel that you were in great danger of being killed?

O No O Yes

10. Are you currently interested in receiving help for a stress, emotional, alcohol or family problem?

O No O Yes

11. Over the LAST 2 WEEKS, how often have you been bothered by any of the following problems?

None	Some	A Lot	
O	O	O	Little interest or pleasure in doing things
O	O	O	Feeling down, depressed, or hopeless
O	O	O	Thoughts that you would be better off dead or hurting yourself in some way

33348

DD FORM 2796, APR 2003

RAND *TR588-AppB.2*

12. Have you ever had any experience that was so frightening, horrible, or upsetting that, IN THE PAST MONTH, you ....

No	Yes	
O	O	Have had any nightmares about it or thought about it when you did not want to?
O	O	Tried hard not to think about it or went out of your way to avoid situations that remind you of it?
O	O	Were constantly on guard, watchful, or easily startled?
O	O	Felt numb or detached from others, activities, or your surroundings?

13. Are you having thoughts or concerns that ...

No	Yes	Unsure	
O	O	O	You may have serious conflicts with your spouse, family members, or close friends?
O	O	O	You might hurt or lose control with someone?

14. While you were deployed, were you exposed to:
(mark all that apply)

No	Sometimes	Often	
O	O	O	DEET insect repellent applied to skin
O	O	O	Pesticide-treated uniforms
O	O	O	Environmental pesticides (like area fogging)
O	O	O	Flea or tick collars
O	O	O	Pesticide strips
O	O	O	Smoke from oil fire
O	O	O	Smoke from burning trash or feces
O	O	O	Vehicle or truck exhaust fumes
O	O	O	Tent heater smoke
O	O	O	JP8 or other fuels
O	O	O	Fog oils (smoke screen)
O	O	O	Solvents
O	O	O	Paints
O	O	O	Ionizing radiation
O	O	O	Radar/microwaves
O	O	O	Lasers
O	O	O	Loud noises
O	O	O	Excessive vibration
O	O	O	Industrial pollution
O	O	O	Sand/dust
O	O	O	Depleted Uranium (If yes, explain)______________________
O	O	O	Other exposures ______________________

15. On how many days did you wear your MOPP over garments? [ ][ ] No. of days

16. How many times did you put on your gas mask because of alerts and NOT because of exercises? [ ][ ] No. of times

17. Were you in or did you enter or closely inspect any destroyed military vehicles?

O No O Yes

18. Do you think you were exposed to any chemical, biological, or radiological warfare agents during this deployment?

O No O Don't know
O Yes, explain with date and location

____

____

____

33348

DD FORM 2796, APR 2003

Health Care Provider Only

SERVICE MEMBER'S SOCIAL SECURITY # ___ — __ — ____

Post-Deployment Health Care Provider Review, Interview, and Assessment

Interview

1. Would you say your health in general is: O Excellent O Very Good O Good O Fair O Poor
2. Do you have any medical or dental problems that developed during this deployment? O Yes O No
3. Are you currently on a profile or light duty? O Yes O No
4. During this deployment have you sought, or do you now intend to seek, counseling or care for your mental health? O Yes O No
5. Do you have concerns about possible exposures or events during this deployment that you feel may affect your health? O Yes O No
   Please list concerns: ____________________
6. Do you currently have any questions or concerns about your health? O Yes O No
   Please list concerns: ____________________

Health Assessment

After my interview/exam of the service member and review of this form, there is a need for further evaluation as indicated below. (More than one may be noted for patients with multiple problems. Further documentation of the problem evaluation to be placed in the service member's medical record.)

REFERRAL INDICATED FOR:

O None
O Cardiac
O Combat/Operational Stress Reaction
O Dental
O Dermatologic
O ENT
O Eye
O Family Problems
O Fatigue, Malaise, Multisystem complaint
O Audiology
O GI
O GU
O GYN
O Mental Health
O Neurologic
O Orthopedic
O Pregnancy
O Pulmonary
O Other ____________________

EXPOSURE CONCERNS (During deployment):

O Environmental
O Occupational
O Combat or mission related
O None

Comments: ____________________

I certify that this review process has been completed.
Provider's signature and stamp:

This visit is coded by V70.5 _ _ 6

Date (dd/mm/yyyy) __ / __ / ____

End of Health Review

33348

DD FORM 2796, APR 2003 ASD(HA) APPROVED

RAND *TR588-AppB.4*

Bibliography

Defense Manpower Data Center, *June 2005 Status of Forces Survey of Reserve Component Members: Administration, Datasets, and Codebook*, DMDC Report No. 2005-023, Washington, D.C., 2005.

DoD—*See* U.S. Department of Defense.

DoL ETA—*See* U.S. Department of Labor, Employment and Training Administration.

Employer Support of the Guard and Reserve, "USERRA: Uniformed Services Employment and Reemployment Rights Act: Ombudsman Services and the Law," Web page, no date. As of July 7, 2008: http://esgr.org/userra.asp

Hoge, Charles W., Jennifer L. Auchterlonie, and Charles S. Milliken, "Mental Health Problems, Use of Mental Health Services, and Attrition from Military Service After Returning from Deployment to Iraq or Afghanistan," *Journal of the American Medical Association*, Vol. 295, 2006, pp. 1023–1032.

Kirchner, Candy, and Mary Montgomery, "UCFE-UCX Programs Federal Claims Control Center Operation," unpublished briefing, Employment and Training Administration, National Interstate Conference, September 6–9, 2005.

Martorell, Francisco, Jacob Alex Klerman, and David S. Loughran, *How Do Earnings Change When Reservists Are Activated? A Reconciliation of Estimates Derived from Survey and Administrative Data*, Santa Monica, Calif.: RAND Corporation, TR-565-OSD, 2008. As of July 3, 2008: http://www.rand.org/pubs/technical_reports/TR565/

National Priorities Project, "Table 1: Educational Attainment of Army Recruits by State, 2005–2007," Web page, no date. As of July 7, 2008: http://www.nationalpriorities.org/table1militaryrecruiting2007

Office of the Chairman of the Joint Chiefs of Staff, "Update Procedures for Deployment Health Surveillance and Readiness," memorandum, MCM-0006-02, February 1, 2002.

Public Law 97-362, Miscellaneous Revenue Act of 1982, October 25, 1982.

Public Law 102-107, Emergency Unemployment Compensation Act of 1991, August 17, 1991.

Public Law 103-353, Uniformed Services Employment and Reemployment Rights Act (USERRA) of 1994, October 13, 1994.

Savych, Bogdan, Jacob Alex Klerman, and David S. Loughran, *Recent Trends in Veteran Unemployment as Measured in the Current Population Survey and the American Community Survey*, Santa Monica, Calif.: RAND Corporation, TR-485-OSD, 2008. As of July 3, 2008: http://www.rand.org/pubs/technical_reports/TR485/

Schwartz, Suzanne S., and Loryn Lancaster, "Comparison of State Unemployment Laws," Web page, 2006. As of July 3, 2008: http://www.workforcesecurity.doleta.gov/unemploy/uilawcompar/2006/comparison2006.asp

Stone, Chad, Robert Greenstein, and Martha Coven, *Addressing Longstanding Gaps in Unemployment Insurance Coverage*, Washington, D.C.: Center for Budget and Policy Priorities, 2007. As of July 3, 2008: http://www.cbpp.org/7-20-07ui.htm

U.S. Code of Federal Regulations, Title 20, Chapter V, Part 614, Unemployment and Compensation for Ex-Servicemembers. As of July 7, 2008:
http://www.dol.gov/dol/allcfr/Title_20/Part_614/toc.htm

U.S. Department of Defense, *Population Representation in the Military Services 2002*, Arlington, Va.: Office of the Secretary of Defense, Personnel and Readiness, 2003.

———, *Population Representation in the Military Services 2005*, Arlington, Va.: Office of the Secretary of Defense, Personnel and Readiness, 2006.

U.S. Department of Labor, Employment and Training Administration, "UI Weekly Claims," Web page, no date. As of July 7, 2008:
http://ows.doleta.gov/unemploy/claims_arch.asp

———, *Handbook No. ETA 384: Unemployment Compensation for Ex-Servicemembers (UCX), 2nd Edition*, Washington, D.C., 1994.

———, "Comparison of State Unemployment Laws," Web page, 2006a. As of July 8, 2008:
http://www.workforcesecurity.doleta.gov/unemploy/uilawcompar/2006/comparison2006.asp

———, "Unemployment Insurance Program Letter No. 27-06," August 2, 2006b. As of July 7, 2008:
http://wdr.doleta.gov/directives/corr_doc.cfm?DOCN=2253

———, "State Law Information," Web page, 2008. As of July 13, 2008:
http://workforcesecurity.doleta.gov/unemploy/statelaws.asp

U.S. Senate, "Jobs for Veterans Act Three Years Later: Are Vets' Employment Programs Working for Veterans?" hearing before the Committee on Veterans Affairs, 109th Congress, 2nd session, S. Hrg. 109-632, February 2, 2006.